MADRUGA MARVEL MEDICAL
BLACK BOOK
SECOND EDITION

Guide to Differential Diagnosis, Mnemonics, and Clinical Pearls

MARIO MADRUGA, MD FACP
FRANCOISE MARVEL, MD

© 2011, 2014 by Madruga & Marvel. All rights reserved.

The text of this publication, or any part thereof, may not be reproduced in any manner whatsoever without written permission from the publisher.

ISBN: 978-1-312-01329-2

Published by Madruga & Marvel, in association with Lulu Publishing Services.

Madruga and Marvel's Medical Black Book: Guide to Differential Diagnosis, Mnemonics, and Clinical Pearls is available in **print** and **eBook** at **www.lulu.com** and **www.amazon.com.** For iPhone, iPod touch, and iPad users, the **M3 Black Book** *app* is available at iTunes App store:
https://itunes.apple.com/us/app/madruga-marvels-medical-black/id465350014?mt=8

For more clinical resources please visit **www.doctechmd.com.**

Cover design & logo art by:
Heather Hoesl
www.h-2design.com

LEGAL DISCLAIMER

This product is intended solely as a reference tool for practicing and experienced healthcare personnel. Laypersons should not rely on any information presented in this book for self-diagnosis or self-treatment, but should ALWAYS consult a physician, go to the nearest emergency room, or contact local emergency medical services for any medical problem. The information in each chapter has been provided by medical professionals, **MADRUGA & MARVEL** have not independently investigated the text supplied by such professionals, however, and cannot guarantee its accuracy. The information contained in this book is not complete and users cannot fully rely on it for treatment or decision-making. Users must therefore rely on their experience and judgment when interpreting any of the information contained herein to an actual patient care situation. In addition, **MADRUGA & MARVEL**, urge users to consult additional reference sources and other healthcare professionals concerning treatments presented herein.

TABLE OF CONTENTS

Chapter 1: Cardiovascular 9
  A: CV Clinical Pearls 9
    AORTIC DISSECTION 9
    ATRIAL FIBRILLATION: $CHA_2 D S_2 VAS_c$ SCORE 9
    ATRIAL MYXOMA 9
    BECK'S TRIAD 10
    CHEST PAIN, CAUSES OF ST ↑ ELEVATION AND ST ↓ DEPRESSION ON EKG 10
    CONGESTIVE HEART FAILURE: OTHER CAUSES FOR ↑ BRAIN NATRIURETIC PEPTIDE 10
    CONGESTIVE HEART FAILURE: ETIOLOGIES [PRIMARY, RESTRICTIVE, ↑ HIGH-OUTPUT] 10
    CORONARY ARTERY DISEASE (CAD) EQUIVALENTS 10
    HYPERTENSION, PREGNANCY 11
    HYPERTENSION, RESISTANT: ETIOLOGIES 11
    LEE REVISED CARDIAC RISK INDEX (LRCRI) 11
    MURMURS: SYSTOLIC TYPES 11
    MYOCARDIAL INFARCTION, MECHANICAL COMPLICATIONS 12
    PERICARDIAL EFFUSION: TOP 5 CAUSES AND OTHERS 12
    R-WAVE: CAUSES OF POOR R-WAVE PROGRESSION 12
    SYNCOPE: ETIOLOGY, HPI, AND WORKUP 12
    SINUS TACHYCARDIA IN THE HOSPITALIZED PATIENT 12
    TACHYCARDIA, WIDE-QRS VS NARROW QRS 13
    TORSADES DE POINTES: CAUSES 13
  B. CV Differential Diagnosis 14
    CHEST PAIN IN THE YOUNG 14
    CAUSES OF CORONARY DISSECTION 14
    GIANT T WAVE INVERSION 14
    HYPOTENSION 14
    JUGULAR VENOUS DISTENSION WITHOUT RALES 14
    MYOCARDIAL INFARCTION, NON-ATHEROSCLEROTIC CAUSES 15
    TROPONINS, COMMON CAUSES OF ELEVATED TROPONINS OTHER THAN ACUTE CORONARY SYNDROMES 15
  C. CV Mnemonics 16
    AORTIC STENOSIS: ***ASD532*** 16
    ACUTE ARTERIAL OCCLUSION: ***6 Ps*** 16
    ATRIAL FIBRILLATION: ***I SMART CHAP*** 16
    CHEST PAIN, ACUTE: ***PS DEATH*** 16
    CONGESTIVE HEART FAILURE: ***FAILURE*** 17
    CORONARY RISK FACTORS: ***CAD HDL*** 17
    HYPERTENSION, SECONDARY CAUSES: ***CCRAPPPS*** 17
    JUGULAR VENOUS PRESSURE: ***PQRST*** 17
    PERICARDITIS, CAUSES OF: ***CARDIAC RINDD*** 18
    SINUS TACHYCARDIA, CAUSES OF: ***TACH FEVER*** 18
  D. CV Clinical Cases 19
    Case #1 19
    Case #2 20-21
Chapter 2: Pulmonary 22
  A: PULM Clinical Pearls 22
    A-A GRADIENT: SIMPLIFIED CALCULATION AND PEARLS 22
    ALLERGIC BRONCHOPULMONARY ASPERGILLOSIS: CLINICAL PICTURE 22
    ASTHMA: SAMTER'S TRIAD 22
    ASTHMA: MIMICKERS AND EXCERBATORS 22
    CHEYNE STOKES RESPIRATIONS: CAUSES 22
    HEMOPTYSIS PEARLS 23
    LOEFFLER'S SYNDROME 23
    PNEUMONIA, ASSOCIATED WITH SPECIAL CONDITIONS 23
    PULMONARY EMBOLISM, PEARLS 23
    PULMONARY EMBOLISM, *WELLS CLINICAL SCORE* FOR DVT 24
    PULMONARY MANIFESTATIONS OF LUPUS 24

- SOLITARY PULMONARY NODULE 24
- **B: PULM Differential Diagnosis 25**
  - HEMOPTYSIS 25
  - MULTIPLE PULMONARY INFILTRATES 25
  - PNEUMONIA: ATYPICAL CAUSES 25
  - PNEUMONIA: UNRESOLVING 25
  - PULMONARY HYPERTENSION 26
  - SINOBRONCHIAL SYNDROMES 26
  - SOLITARY PULMONARY NODULE 26
- **C: PULM Mnemonics 27**
  - ASTHMA EXACERBATION: ***DIPLOMAT*** 27
  - CHRONIC COUGH: ***TB COUGH*** 27
  - HEMOPTYSIS: ***CAVITATES*** 27
  - LUNG CANCER, METASTASIS: ***BLAB*** 27
  - PULMONARY EDEMA: ***NOT CARDIAC*** 28
  - PULMONARY FIBROSIS: ***SCAR*** 28
  - SHORTNESS OF BREATH: ***CDSPIES*** 28
  - TUBERCULOSIS, MULTIDRUG REGIMEN: ***RIPE*** 28
- **D: PULM Clinical Case 29**
  - Case #1 29
  - Case #2 30
- **Chapter 3: Gastroenterology 31**
  - **A. GI Clinical Pearls 31**
    - CHARCOT TRIAD 31
    - CONSTIPATION: CAUSES, WORK-UP, TREATMENT, PREVENTION 31
    - DIARRHEA: LIKELY SOURCES IN HIV PATIENTS 31
    - DIARRHEA: TYPES, PEARLS, MANAGEMENT 32
    - DISCRIMINANT FUNCTION (ALCOHOLIC HEPATITIS) 32
    - FOOD POISONING SYNDROME, TOXINS 32
    - GASTROINTESTINAL BLEED, LOWER: PEARLS 33
    - GASTROINTESTINAL BLEED, UPPER: HPI PERTINENT POSITIVES 33
    - HEPATITIS B: SERUM MARKERS 33
    - HEPATITIS C: EXTRA-HEPATIC MANIFESTATIONS 33
    - LIVER FUNCTION TESTS, PATTERNS 34
    - LIVER, TOXIC OTC PEARLS 35
    - NAUSEA AND VOMITING, INTRACTABLE: WORKUP 35
    - PANCREATITIS 35
    - PUD: ENDOSCOPIC CHARACTERISTICS OF PUD UPPER-GI BLEEDING AND RATES OF REBLEEDING 35
  - **B. GI Differential Diagnosis 36**
    - ASCITES, DDX (BASED ON SAAG) 36
    - CRAMPING DISORDERS 36
    - GASTRIC FOLDS, THICKENED 36
    - GASTRIC LESIONS, MULTIPLE 36
    - GASTROINTESTINAL BLEED, UPPER 36
    - GASTROINTESTINAL BLEED, LOWER 37
    - NAUSEA AND VOMITING 37
    - PANCREATITIS 37
    - PANCREATITIS CAUSED BY MEDICATIONS 37
    - PORTAL VEIN THROMBOSIS, RISK FACTORS 38
    - SPLENOMEGALY, MASSIVE 38
  - **C. GI Mnemonics 39**
    - ABDOMINAL PAIN: NON-APPENDICEAL RLQ PAIN: ***APPENDICITIS IS NOT THE ONLY CAUSE OF RLQ PAIN*** 39
    - CHOLELITHIASIS: ***5 Fs*** 39
    - HEPATITIS, CAUSES: ***ABCDE*** 39
    - HEPATITIS C: ***5 DEUCES OF HEPATITIS C*** 39
    - HEPATITIS, DRUG-INDUCED OR ACUTE INTERSTITIAL NEPHRITIS: ***FARE*** 40

HEPATOCELLULAR CANCER RISK FACTORS: ***WATCH FOR ABC***	40
ILEUS: ***5 Ps***	40
MECKEL'S DIVERTICULUM: ***RULE OF 2s***	40
NEEDLE STICK ACCIDENTS: ***3s OF NEEDLE STICK RISK***	40
NON-ALCOHOLIC FATTY LIVER DISEASE: ***NAFLD-DROP***	40
PANCREATITIS: ***GET SMASHED***	41
VOMITING EXTRA-GI CAUSES: ***VOMITTING***	41
Chapter 4: Renal & Genitourinary	**42**
A. RENAL/GU Clinical Pearls	**42**
ACUTE RENAL FAILURE: PRERENAL, INTRARENAL, POST-RENAL	42
CHRONIC KIDNEY DISEASE STAGES	42
CHRONIC RENAL FAILURE, TOP 5 CAUSES	42
CHRONIC RENAL FAILURE PATIENTS, "GOLDEN RULES"	42
CONTRAST-INDUCED NEPHROPATHY	43
DIALYSIS: ABSOLUTE INDICATIONS	43
NEPHROTIC SYNDROMES, CRITERIA, AND CLINICAL PICTURE	43
NEPHROTIC SYNDROME: WORKUP	43
RENAL FAILURE, BONE DISEASES	43
RENAL ARTERY STENOSIS: CLINICAL CLUES	44
URINARY RETENTION, POST-OPERATIVE	44
B. RENAL/GU Differential Diagnosis	**45**
GLOMERULAR DISEASES IN PATIENTS WITH HIV/HEP C	45
NEPHROTIC SYNDROME: PRIMARY RENAL ETIOLOGIES VS SECONDARY CAUSES	45
RENAL CYSTS	45
URINARY RETENTION, POST-OPERATIVE	45
C. RENAL/GU Mnemonics	**46**
ACUTE RENAL FAILURE: ***CORPORATE VICE PRESIDENTS HATE DOGS MATING***	46
HEMATURIA: ***INEPT GUN***	46
NEPHROTIC SYNDROME: ***THIS LAD HAS NEPHROTIC SYNDROME***	46
URINARY TRACT INFECTIONS: ***SEEKS PP***	47
Chapter 5: Neurology	**48**
A. NEURO Clinical Pearls	**48**
ENCEPHALOPATHY, HEPATIC: PRECIPITATING FACTORS	48
KORSAKOFF DEMENTIA	48
NEURO TRACTS: CROSSING AND COLUMNS	48
SEIZURE – EPILEPSY – STATUS EPILEPTICUS	48
SPINAL CORD INJURIES: CLINICAL MANAGEMENT	49
STROKE, CAUSES OF STROKE OTHER THAN ATHEROSCLEROSIS OR EMBOLI	49
STROKE, CLINICAL LACUNAR SYNDROME AND INFARCT LOCATION	49
WERNICKE'S TRIAD	49
B. NEURO Differential Diagnosis	**50**
CEREBELLAR ATAXIA	50
FACIAL NERVE PALSY	50
PERIPHERAL NEUROPATHY: PAINFUL	50
STROKE IN THE YOUNG	51
TRANSVERSE MYELITIS	51
C. NEURO Mnemonics	**52**
ALTERED MENTAL STATUS: ***DELERIUMS***	52
DEMENTIA ETIOLOGIES: ***DEMENTIASS***	52
HORNER'S SYNDROME: ***AMP***	52
NORMAL PRESSURE HYDROCEPHALUS: ***3Ws***	52
PARKINSON'S DISEASE SIGNS: ***SMART***	52
PERIPHERAL NEUROPATHY: ***Dang Therapist***	53
SUBARACHNOID HEMORRHAGE: ***BAT CAVES***	53
STROKE: RISK FACTORS: ***HEADACHES***	54
STROKE: MIDDLE CEREBRAL ARTERY: ***CHANGE***	54
STROKE: POSTERIOR CIRCULATION: ***4 Deadly Ds***	54
STROKE: CAUSES IN YOUNG PATIENTS: ***7Cs***	54

Chapter 6: Infectious Diseases ... 55
A. ID Clinical Pearls ... 55
- ANTIBIOTICS: SIDE EFFECTS ... 55
- CLOSTRIDIUM DIFFICILE: NEW STRAIN ... 55
- EHRLICHIOSIS, CLINICAL PICTURE ... 55
- ENTEROCOCCUS: SOURCES OF INFECTION AND TREATMENT ... 55
- FEVER OF UNKNOWN ORIGIN: DDx, PEARLS, WORK-UP ... 56
- FEVER, CAUSES OF FEVER IN INTENSIVE CARE UNIT ... 57
- FEVER, *LAST DITCH* WORKUPS TO VERIFY A CAUSE ... 57
- LEMIERRE'S SYNDROME ... 57
- MENINGITIS AND ENCEPHALITIS: ETIOLOGIES ... 57
- MENINGITIS: CEREBROSPINAL FLUID ANALYSIS WORKUP ... 57
- NOSOCOMIAL INFECTION, *5 BUGS TO KEEP IN MIND* ... 57
- SEPSIS: Hx, ETIOLOGY, GRAM NEGATIVES ... 58
- WEST NILE VIRUS: SIGNS AND SYMPTOMS ... 58

B. ID Differential Diagnosis ... 59
- BACTEREMIA, GRAM NEGATIVE ... 59
- HIV LYMPHADENOPATHY/PULMONARY INFILTRATES/CYTOPENIAS ... 59
- PERIPHERAL LYMPHADENOPATHY ... 59
- MENINGITIS AND ENCEPHALITIS, HISTORICAL CLUES TO ETIOLOGY ... 59

C. ID Mnemonics ... 60
- ENDOCARDITIS: *HACEK BB* ... 60
- POST-OPERATIVE FEVER: *5 Ws OF POST-OPERATIVE FEVER* ... 60

D. ID Clinical Cases ... 61
- Case #1 ... 61
- Case #2 ... 62-63
- Case #3 ... 63

Chapter 7: Endocrine & Metabolism ... 64
A. ENDO/METABO Clinical Pearls ... 64
- ADRENAL INCIDENTALOMA: PEARLS AND WORKUP ... 64
- CUSHING'S SYNDROME ... 64
- DIABETES MELLITUS TYPE 2 MEDICATIONS: BLACK BOX WARNING ... 64
- HYPERGLYCEMIA ↑ BLOOD GLUCOSE: SECONDARY CAUSES ... 65
- HYPERKALEMIA ↑ K$^+$: MAJOR CAUSES ... 65
- HYPERTHYROIDISM: CAUSES & PEARLS ... 65
- HYPOCALCEMIA, CAUSES OF ↓ CA^{2+} ... 65
- LACTIC ACIDOSIS ... 66
- MEN SYNDROMES (WERMER'S SYNDROME AND SIPPLE'S SYNDROME) ... 66
- THYROID UPTAKE AND SCAN RESULTS ... 66
- THYROID MEDICATION, PEARLS ... 66
- WHIPPLE'S TRIAD ... 66

B. ENDO/METABO Differential Diagnosis ... 67
- HYPOGLYCEMIA ... 67
- HYPOKALEMIA, METABOLIC ALKALOSIS, AND HYPERTENSION ... 67
- HYPONATREMIC, EUVOLEMIC ... 67
- HYPOPHOSPHATEMIA ... 67

C. ENDO/METABO Mnemonics ... 68
- ACIDOSIS, ANION GAP: *A MUD PILES* ... 68
- HYPERCALCEMIA, COMPLICATIONS: *STONES, BONES, ABDOMINAL MOANS* ... 68
- HYPERCALCEMIA, MAJOR CAUSES: *MISHAP* ... 68
- HYPOGLYCEMIA, MAJOR CAUSES: **NOT ONLY NUTRITION EXPLAINS IT** ... 69
- PHEOCHROMOCYTOMA: *5Ps AND RULE OF 10s* ... 69

D. ENDO/METABO Clinical Cases ... 70
- Case #1 ... 70-71

Chapter 8: Hematolgy & Oncology ... 72
A. HEME/ONC Clinical Pearls ... 72
- ALPHA & BETA THALASSEMIA ... 72
- ANEMIA, MICROCYTIC: (MCV<80 FL) ... 72

ANEMIA, MACROCYTIC 73
ANEMIA: KEY LAB FINDINGS & PERIPHERAL BLOOD SMEAR ANALYSIS W/SUPPORTING
DIAGNOSIS 72
CHEMOTHERAPEUTIC DRUGS: MAJOR SIDE EFFECTS 73
COAGULATION PEARLS: ZEBRAS 73
SICKLE CELL DISEASE: CLINICAL MANIFESTATIONS 73
PARANEOPLASTIC SYNDROMES, ASSOCIATED WITH SMALL CELL LUNG CANCER (SCLC) 73
VIRCHOW TRIAD 73
B. HEME/ONC Differential Diagnosis **74**
ANEMIA OF CHRONIC DISEASE 74
ATYPICAL LYMPHOCYTES 74
EOSINOPHILIA 74
IGE ELEVATION 74
HEMOLYTIC ANEMIA, CAUSES 75
LYMPHOCYTIC PLEOCYTOSIS 75
MONOCLONAL: GAMMOPATHY 75
THROMBOCYTOSIS, REACTIVE 75
C. HEME/ONC Mnemonics **76**
ANEMIA, MACROCYTIC: *ABCDEF* 76
ANEMIA, MICROCYTIC: *TICS* 76
DISSEMINATED INTRAVASCULAR COAGULATION: *DISSEMINATED* 76
EOSINOPHILIA: *NAACP* 76
METASTATIC DISEASE: *KIND OF TUMORS LEAPING TO BONE/MA-PA-PB-KTL* 77
TARGET CELLS DIFFERENTIAL: *HALT* 77
THROMBOTIC THROMBOCYTOPENIC PURPURA: *FAT RUN* 77
THROMBOCYTOPENIA: *VIC G. SAID TO TED AND SAMMM* 78

Chapter 9: Rheumatology 79
A. RHEUM Clinical Pearls **79**
ANTIBODIES (SERUM) FOR DIAGNOSIS 79
ANTIPHOSPHOLIPID SYNDROME: CRITERIA FOR THE DEFINITE DIAGNOSIS 79
ERYTHEMA NODOSUM: CAUSES 80
PROXIMAL MUSCLE WEAKNESS, PEARLS 80
SYSTEMIC INFLAMMATORY CONDITIONS ASSOCIATED WITH CARDIOVASCULAR DISEASE 80
UVEITIS, ASSOCIATED WITH RHEUMATIC DISEASE 80
VASCULITIS 81
B. RHEUM Differential Diagnosis **82**
ARTHRITIS, MONOARTICULAR 82
ARTHRITIS, POLYARTICULAR 82
FATIGUE, MUSCULAR WEAKNESS 82
EVAN'S SYNDROME: SECONDARY CAUSES 82
C. RHEUM Mnemonics **83**
COLD AGGLUTININ: *5 MS* 83
CALCIUM PYROPHOSPHATE DEPOSITION DISEASE: *5 HS* 83
LUPUS: *SOAP BRAIN MD* 83
RHEUMATOID ARTHRITIS: *RHEUMATISM* 83
D. RHEUM Clinical Cases **84**
Case #1 84-85
Case #2 86
Case #3 87
Case #4 88
Case #5 89
Case #6 90

Chapter 10: Dermatology 91
A. DERM Clinical Pearls **91**
HIV, DERMATOLOGICAL MANIFESTATIONS 91
ORAL ULCERS: PEARLS AND DDX 91
URTICARIA WORK-UP 91
B. DERM Differential Diagnosis **92**

Erythroderma, Diffuse	92
Pruritus	92
Rash, Papulosquamous	92
C. DERM Mnemonics	**93**
Erythema Nodosum: Associated Conditions: ***SPUD BITS***	93
Melanoma/Malignancy: ***ABCDE***	93
Ulcers, Genital: ***Some Girls...but Fellows***	93
White Patch On Skin: ***VITILIGO PATCH***	93
Chapter 11: Pharmacology	**94**
A. PHARM Clinical Pearls	**94**
Amiodarone: 5 Classic Side Effects	94
Anticholinergic Overdose	94
Lasix Drip Management	94
Salicylate (Aspirin) Toxicity	94
Warfarin: Drugs that Potentiate Warfarin's Effect	95
Warfarin, Managing the Overcoagulated Patient	95
Warfarin's Effect on Hypercoag Tests	95
Warfarin, Diet Guidelines and Restrictions	96
B. PHARM Mnemonics	**96**
Anticholinergic/TCAs/Benadryl Overdose: ***Hot, Blind, Mad***	96
Chapter 12: General	**97**
A. GEN Clinical Pearls	**97**
Amyloidosis: Clinical Picture	97
Beri Beri: Clinical Picture	97
Creatinine Phosphokinase (CPK): Causes of Elevation	97
Edema: Causes of Edema in Lower Extremities	97
Hair Loss: Telogen Effluvium	97
Increased Intracranial Pressure: Non Structural Causes	97
IV Immune Globulin: Conditions that may warrant administration of IVIG	98
Löfgren syndrome	98
Overdose and Antidotes	98
Sedimentation Rate	99
Unintentional Weight Loss, Causes	99
B. GEN Differential Diagnosis	**99**
Complement States, Decreased	99
Weakness	99
C. GEN Mnemonics	**100**
Back Pain: ***DISK MASS***	100
Diagnostic Approach to Challenging Cases: ***I VINDICATE***	100
NOTES	**101-105**

CHAPTER 1: CARDIOVASCULAR

A. CV Clinical Pearls

Aortic Dissection	
Risk Factors - ABCs	- Atheroscl. / Aging/ Aortic aneu./Aortic coarc. - BP↑/ **B**aby (pregnancy) /**B**icuspid aortic valve - **C**onnective Tissue Disorders, **C**ystic medial necr. **Genetic disorders** with ↑ risk for Aortic Dissection: → Turner, Marfan, Ehlers-Danlos, Loeys-Dietz
Clinical Picture	- Sudden **"tearing"** chest / back pain - PE: **interarm BP > 20 mm Hg differential.** - **Widened mediastinum** on Chest X-Ray
1st Steps in Management	- **Blood Pressure control,** Esmolol and Nitroprusside - ↓Heart rate ↓Afterload reduction to ↓shear forces.
Work-up	☑ Spiral CT ☑ MRI ☑ Trans-esophageal echocardiogram
Complications	- Cardiac Tamponade, Stroke, Myocardial Infarction **Pearl:** if the dissection goes to the coronary arteries, the **right coronary artery** is the most commonly affected.

Atrial Fibrillation Stroke Risk: CHA$_2$DS$_2$VAS$_c$ Score		
	Condition	Points
C	**C**ongestive heart failure	1
H	**H**ypertension	1
A$_2$	**A**ge>75	2
D	**D**iabetes mellitus	1
S$_2$	Prior **S**troke or TIA	2
V	**V**ascular disease (PAD, MI, Aort. plaq.)	1
A	Age 65-74	1
S$_c$	**S**ex category (female w/ risk factors)	1
Aspirin [81-325 mg] daily indicated.		Score of **0-1**
Warfarin* indicated for patient.		Score **2 or higher**

*ASA + Plavix acceptable in patients in which Warfarin contraindicated.

Atrial Myxoma

- Most common **1° intracardiac tumor**, usually **LEFT** atrium.
- Sx: Fever/wt loss/neuro sym 2° → **tumor embolization**
- Dx: **Mass on 2D-echocardiography**, ↑Sedimentation Rate, ↑CRP, ↑interleukin-6 levels, anemia of chronic disease.

Beck's Triad	
1. Jugular venous distension 2. Muffled heart sounds 3. Hypotension	Triad suggests diagnosis of: **Cardiac tamponade**

Chest Pain, Causes of ST ↑ Elevation and ↓ Depression on EKG	
ST Segment ↑ Elevation	ST-Segment ↓ Depression

• Ischemia, Infarct
• Brugada
• Early repolarization
  (ex. J-Point elevation)
• Injury (ex. contusion)
• Prinzmetal's angina
  (ex. coronary vasospasm)
• Takotsubo cardiomyopathy
• Ventricular aneurysm
	o Left ventricular hypertrophy w/ strain
	o Digoxin toxicity
	o Reciprocal ST-depression
	o Subendocardial ischemia

Congestive Heart Failure: Other Causes for Elevated Brain Natriuretic Peptide (BNP)

- Cor pulmonale
- Pulmonary embolism
- Pulmonary hypertension
- Renal failure
- Restrictive cardiomyopathy

Congestive Heart Failure: Etiologies		
Primary	• Aortic stenosis • Cardiomyopathy • Cor pulmonale ("CHF Mimic") • CAD / Ischemia • Diastolic dysfunct. • Endocarditis • Hypertension	• Idiopathic hypertrophic subaortic stenosis • Myocarditis • Post-partum • Substances (ex. Alcohol, Cocaine) • Valvular heart dz
Restrictive/ Infiltrative	• Amyloidosis • Hemochromatosis	• Radiation • Sarcoid
↑High Output	• Anemia • Arteriovenous shunt • Beri Beri	• Cirrhosis • Hyperthyroid • Paget's disease • Sepsis

Coronary Artery Disease (CAD) Equivalents

- Diabetes mellitus
- Peripheral artery disease
- Abdominal aortic aneurysm

Hypertension, Pregnancy

Take Home Message: Only certain medications have been approved for treatment of hypertension during pregnancy.

- ☑ Labetalol
- ☑ Methyldopa
- ☑ Hydralazine
- ☑ Procardia

Hypertension, Resistant: Etiologies

Drugs, Toxins	• Alcohol • Birth control pills • Cyclosporine • Licorice • NSAIDs (COX-1, COX-2 Inhibitors) • Steroids	Stimulants / Herbals / Appetite ↓ / Decongest.: ○ Cocaine, Ephedra ○ Bitter Orange, Guarana, Gotu Kola, Yerba Mate, Ma huang ○ Pseudoephedrine
Renal	• Erythropoietin • Renal Artery Stenosis	• Glomerulonephritis
Other Causes	- Endocrine causes [Pheochromocytoma / Aldosterone] - Noncompliance with anti-hypertensive meds. - Sleep apnea - Parathyroid also Thyroid dysfunction (test TSH) - Cushing's - Coarctation of the Aorta	

Lee Revised Cardiac Risk Index (LRCRI)

Points	Description
1	High risk surgery
1	History of coronary artery disease
1	History of cerebrovascular accident
1	History of congestive heart failure
1	Diabetes mellitus (requiring insulin)
1	Chronic kidney disease (Cr>2)

Score Interpretation Risk of Perioperative Cardiac Complications

0-1	Low Risk 1%
2	Intermediate 2.5%
3 or higher	High Risk 5%

Murmurs: Systolic Types

- Aortic stenosis
- Atrial septal defect
- Ventricular septal defect
- Mitral regurgitation
- Mitral valve prolapse
- Idiopathic hypertrophic subaortic stenosis
- Pulmonary stenosis
- Tricuspid regurgitation

Myocardial Infarction, Mechanical Complications		
Myocardial Infarction Type	**Mech. Complication**	**Other**
- Left anterior descending - Most occur within 4 days. - Left Anterior Descending /Anteroseptal MI - ↑PAP, O2 step up with pulmonary artery catheterization - Right Coronary Artery - Post. papillary mscl. rupture	- LV free wall rupture - VSD - Mitral regurg.	- Mural Thrombus - Left Ventricular Aneurysm

Pericardial Effusion: Top 5 Causes and Others	
1. Viral (CMV, coxsackie, echo, HIV) 2. Tuberculosis /Fungal 3. Malignancy 4. Systemic lupus erythematosus 5. Bacterial	**Other Causes:** - Aortic Dissection, Cocaine, Hypothyroid, Sarcoid, Uremia - Post-MI (Dressler's syndrome) - Meds (Minoxidil, Procainamide)

R-Wave: Causes of Poor R-Wave Progression	
- Anterior Wall MI [old] - Cardiomyopathy - Cardiac Tamponade	- COPD - Lead placement - Obesity

Syncope: Etiology, HPI, and Workup	
HPI Positive+	- Prodrome - What were you doing? - Post-ictal - Chest pain
Etiology	- Alcohol - Arrhythmia - Carotid sinus sensitivity - Dehydration - GI bleed - Medications/Iatrogenic - Neurocardiogenic syncope - Vasovagal *Pearl: Important to **rule-out deadly causes of syncope** – ex. structural heart disease, Aortic stenosis, Idiopathic hypertrophic subaortic stenosis, Wolff–Parkinson–White syndrome, Pulmonary Embolism, QT-prolongation, Brugada syndrome.
Workup Level 1	☑ Telemetry ☑ ECHO ☑ ABG ☑ CBC, CMP, UA, UDP, EtOH ☑ CXR PA/LAT ☑ Orthostatics ☑ EKG – WPW – QT Prolongation/Brugada (Rule Out) ☑ CT (focal neuro deficits)
Workup Level 2	☑ Stress test ☑ Tilt table test ☑ EPS ☑ Loop Recorder or event monitor ☑ EEG/MRI if seizure is thought to be a possibility

Sinus Tachycardia in the Hospitalized Patient

TOP 6 Causes of <u>New</u> Onset Tachycardia.

NOTE: Sinus tachycardia = **sinus rhythm** with a **rate > 100 bpm**.

1. Pain
2. Pulmonary embolism
3. Hyperthyroid
4. Medications
5. Anemia or Hypovolemia
6. Sepsis

Tachycardia, Wide-QRS vs Narrow QRS

Wide QRS	Narrow QRS
- Bundle branch block - Supraventricular tachycardia w/ Aberrancy - Torsades de Pointes - Ventricular tachycardia - Wolf-Parkinson-White (assoc/w Atrial fibrillation)	o Atrial flutter o Supraventricular tachycardia o Atrial fibrillation o Multifocal atrial tachycardia

Torsades de Pointes: Causes

- Brugada syndrome
- Jervell & Lange-Nielsen syndrome
  - Congenitally long QT & **deafness**
- Decreased electrolytes
  - Hypokalemia ↓K^+
  - Hypomagnesemia ↓Mg^{2+}
- Medications (including but not limited to the following meds):
  - Amiodarone
  - Disopyramide
  - Encainide
  - Flecainide
  - Procainamide
  - Quinidine
  - Sotalol
- Romano Ward syndrome (i.e., long QT)
- Takotsubo cardiomyopathy

B. CV Differential Diagnosis

Chest Pain in the Young: DDx

- Bacterial endocarditis
- Cocaine
- Congenital (ex. Anomalous coronary arteries)
- Familial hypercholesterolemia
- Hypercoagulable state
- Idiopathic hypertrophic subaortic stenosis (IHSS)
- Kawasaki's disease as an infant (ex. Coronary artery aneurysms)
- Vasculitis (ex. Lupus)
- Vasospasm

Coronary Dissection, Causes: DDx

- Cathertization +/- PCI
- CABG
- Chest wall trauma
- Extension of aortic dissection
- Peripartum or hyperestrogenism states
- **Spontaneous**
  - PAN, SLE, Churg-Strauss
  - Sarcoid
  - Kawasaki disease
  - Marfans
  - Fibromuscular dysplasia
  - Ehlers-Danlos Type IV

Giant T Wave Inversion: DDx

- Acute abdomen (Acute Panc.)
- Cocaine
- Complete heart block
- Elevated intracranial pressure
- Hypertrophic obstructive cardiomyopathy (Apical)
- Subarachnoid hemorrhage
- Non-Q MI
- Severe RVH
- Post-pacemaker syndrome
- Wolff-Parkinson-White syndrome

Hypotension: DDx

- Adrenal insufficiency
- Anaphylaxis
- Anemia – Bleed
- Anesthesia – Spinal
- Benzodiazepines, Sedatives
- Myocardial infarction
- Narcotics
- Pancreatitis
- Peritonitis
- Pulmonary embolism
- Post-cardiac catheterization
  - Groin bleed, Dissection, Fistula, Emboli, Retroperitoneal bleed
- Sepsis – Fever
- Vomiting and diarrhea

Jugular Venous Distension WITHOUT Rales: DDx

- Cardiac tamponade
- Constrictive pericarditis
- COPD
- Cor pulmonale
- Pulmonary embolism
- Right ventricular MI
- Tension pneumothorax

Myocardial Infarction, Non-Atherosclerotic Causes: DDx

Arteritis	<ul><li>Kawasaki syndrome</li><li>Takayasu arteritis</li><li>Polyarteritis nodosa</li></ul>	<ul><li>Churg Strauss syndrome</li><li>Syphilis</li></ul>
Collagen Vascular Dz	<ul><li>SLE</li><li>Systemic sclerosis</li></ul>	<ul><li>Mixed connective tissue disease</li></ul>
Mural Thickening ----OR----- Intimal Proliferation	<ul><li>Amyloidosis</li><li>Cocaine</li><li>Fabry disease</li><li>Homocystinemia</li><li>Inherited and acquired hypercoagulable states</li></ul>	<ul><li>Prinzmetal angina</li><li>Pseudoxanthoma elasticum</li><li>Radiation treatment</li></ul>
Others	Spontaneous coronary artery dissection	

Troponins, Common Causes of Elevated Troponins OTHER than Acute Coronary Syndromes: DDx

Cardiovascular	<ul><li>Aortic dissection</li><li>Arrhythmias</li><li>Cardiac infiltrative d/o</li><li>Cardiomyopathy</li><li>Congestive heart failure</li><li>Endocarditis</li><li>Pericarditis/Myocarditis</li></ul>	<ul><li>Cardiac contusion</li><li>Stroke</li><li>SAH</li><li>ICH</li><li>Kawasaki disease</li><li>Takotsubo</li></ul>
Respiratory	<ul><li>ARDS</li></ul>	<ul><li>Pulmonary embolism</li></ul>
Chronic Diseases	<ul><li>ESRD</li><li>Neurofibromatosis</li></ul>	<ul><li>Duchenne MD</li><li>HTN</li></ul>
False Pos (+) Interference	<ul><li>Rheumatoid factor</li><li>Heterophile antibodies</li></ul>	<ul><li>Alkaline phosphatase</li><li>Hemolysis</li></ul>
Other	Sepsis	

C. CV Mnemonics

Aortic Stenosis

A	**A**ngina
S	**S**yncope
D	**D**yspnea (CHF)
5	**5** year prognosis for **A**ngina
3	**3** year prognosis for **S**yncope
2	**2** year prognosis for **D**yspnea (CHF)

ASD – 532
Aortic stenosis symptoms and prognosis **WITHOUT corrective surgery**.

Acute Arterial Occlusion

P	**P**ain
P	**P**allor
P	**P**ulselessness
P	**P**aralysis
P	**P**oikilothermia
P	**P**aresthesias

6 Ps
Signs and symptoms of **acute arterial occlusion** (ex. thrombi, emboli).
**MEDICAL EMERGENCY = clinical signs of impending tissue loss. Early diagnosis and intervention is critical.

Atrial Fibrillation

I	**I**nflammatory (Pericarditis/Myocarditis)
S	**S**urgery (Post-CABG)
M	**M**edications
A	**A**therosclerotic coronary disease
R	**R**heumatic valvular disease
T	**T**hyroid
C	**C**ongenital heart disease
H	**H**ypertension
A	**A**lcohol
P	**P**ulmonary Disease* – PE/O2 ↓

I SMART CHAP
Main causes of **Atrial fibrillation**.

Other causes:
Cardiomyopathy, Sepsis, MI, Cocaine, Theophylline, Anemia, *PE, *Hypoxemia [ex. Asthma, COPD], Idiopathic ("Lone AF")

Chest Pain, Acute

P	**P**neumothorax
S	**S**epsis (Pneumonia)
D	**D**issection
E	**E**mbolism
A	**A**ngina
T	**T**amponade
H	**H**eart attack

PS Death
Differential diagnoses for **deadly** causes of acute **Chest Pain**.

Non-Lethal causes of CP:
MS – HSV, Anxiety, GERD PUD (spasm) Esophagitis, Biliary – Stones, Pancreatitis, Pericarditis / Myocarditis.

Congestive Heart Failure

F	Forgot medication/Fluid retention	**FAILURE**
A	Arrhythmia/Anemia	**Heart failure** exacerbators.
I	Ischemia/Infarction/Infection	
L	Lifestyle: ↑High Salt diet	***Other Causes:**
U	Upregulation of ↑Cardiac Output: pregnancy, hyperthyroidism, HTN	HTN, worsening valve, Sepsis, GI bleed /Anemia, Fluid Retention States,
R	Renal failure	Meds (Glucophage,
E	Embolism (pulmonary)	Avandia, Actos).

Coronary Artery Disease

C	Cigarettes	**CAD HDL**
A	Age (M > 45 F> 55 yo)	Top 6 **risk factors** for
D	Diabetes mellitus*	**coronary artery disease.**
H	Hypertension	
D	Death from MI in family history	***Greatest** risk factor:
L	↑High LDL (>100), Low HDL (<35)	**Diabetes Mellitus**

2° Hypertension

C	Coarctation of the aorta	**CCRAPPPS**
C	Cushing's or Steroids	Causes of **Secondary Hypertension.**
R	Renal artery stenosis	***Fluoxetine, lithium,**
A	Aldosteronism	TCAs, Naproxen,
P	Pheochromocytoma	ibuprofen,
P	Pill (Estrogen OCPs), other drug*	decongestants, cocaine,
P	Parathyroid, Thyroid dysfunction	ephedra, ginseng, ma
S	Sleep Apnea	huang.

Jugular Venous Pressure

P	Pericarditis (constrictive, effusion)	**PQRST**
Q	↑Quantity of Fluid (overload)	Raised **JVP** differential.
R	Right Heart Failure	
S	Superior vena caval obstruction	**Pearl** – Examine right IJV, 45° angle, tangential light.
T	Tamponade (cardiac)	See abnormalities below.
	Tricuspid stenosis or regurg	

Pearls: JVP Abnormalities and Likely Diagnosis

A-wave [NO A-wave = A-FIB]	▪ Large a waves → right ventricular hypertrophy ▪ Cannon waves→ complete heart block and VTach
Prominent V-wave	▪ Tricuspid regurgitation
Slow y descent	▪ Tricuspid stenosis ▪ Right atrial myxoma
Steep y decent [Rapid rise/fall JVP = Friedreich's sign]	▪ Right ventricular failure ▪ Constrictive pericarditis ▪ Tricuspid regurgitation

Pericarditis

C	Collagen vascular disease	
A	Autoimmune	
R	Radiation (breast and lung CA)	
D	Drugs*	
I	Infections**	
A	Acute renal failure	
C	Cardiac infarction	
R	Rheumatic fever	
I	Injury (trauma, postpericardiotomy)	
N	Neoplasms (breast, lung, Hodgkin's)	
D	Dressler's syndrome (*post–MI syndr.*)	
D	Dissection (aortic dissection)	

CARDIAC RINDD
Causes of **Pericarditis**.

*Drugs - hydralazine, doxorubicin, INH, penicillin, Rifampin.

**Infx: Coxsack, Hep, HIV, Flu, Mumps, Varicella, Measles, Gram+/-, Fungi – blasto, candida, histo, *Echinococcus granulosus*

Sinus Tachycardia

T	Tamponade/Thyrotoxicosis	
A	Anemia/Anxiety	
C	CHF/Chronic pulmonary disease	
H	Hypotension (and shock)/Hypoxia	
F	Fever/Flight or Fight Response	
E	Excruciating Pain	
V	Volume depletion	
E	Exercise (catecholamines)	
Rx	Stimulants, nicotine, caff, drugs, epi	

TACH FEVER
Causes of **sinus tachycardia**.

Others:
Pheochromocytoma, Sepsis, Pulmonary embolism, Acute coronary ischemia and MI.

D. CV Clinical Case #1

History:
- 65-yo male presents with abdominal and leg swelling with clear lungs, no murmur, no displaced PMI, no jaundice, no spiders, no history of alcoholism, no history of hepatitis or hepatitis risk factors.

Physical Exam:
- Remarkable for **jugular venous distension** and **lower extremity edema.**

Assessment:
1. Consider the differential diagnosis for **JVD with NO Rales**:

- Pulmonary embolism
- Cor pulmonale
- Tamponade (*Pearl: depends on chronicity)
- Constrictive pericarditis
- Right ventricular infarct
- Restrictive cardiomyop.

2. Review the clinical pearl for **Causes of Edema in Lower Extremities**:

Cardiac	<ul><li>CHF</li><li>Constrictive pericarditis</li><li>Cor pulmonale</li><li>Inferior vena cava syn.</li></ul> Hypertension Pulmonary HTN ↑High-output fail Valvular dysfunction
Liver	↓ Low albumin
Lymphatics	Lymph blockage
Malnutrition	Hypoalbuminemia 2° to malnutrition.
Medications	Calcium channel blockers Estrogen Steroids Minoxidil NSAIDs
Renal	Nephrotic syndrome
Thyroid	Hypothyroidism

Final Diagnosis: Constrictive pericarditis
- Pericardium **rigid, fibrotic, adheres** to the myocardium, ↓ fxn
  - ☑ **Hepatomegaly w/ ascites** and **lower extremity edema**
  - ☑ **Raised JVP** with positive Kussmaul sign
  - ☑ **Pulsus paradoxus** (but usually less severe than in tamponade)
  - ☑ **Pericardial knock** (high-pitched early diastolic added sound)
  - ☑ **Calcification visible** on the **lateral CXR**

Pearls: Constrictive Pericarditis
- **Clinical pearl** – this is ascites and lower extremity edema in a patient without liver disease, which makes this a **non-hepatic ascites**.
- **Physical Diagnosis Skills** – key PE finding is **jugular venous distension**. **Pearl** – Examine right IJV, 45° angle, tangential light.

D. CV Clinical Case #2

History:
- 62-yo male c/o rapid and irregular heartbeat he describes as "fluttering in his chest" that lasts one hour and then resolves. Pt admits it has occurred on and off for last 1 month. ROS negative for CP, nausea, diaphoresis. Pt admits to fatigue and nighttime snoring. Pt has history of GERD, but no cardiac or anxiety disorders. Drinks 1-2 glasses of wine per week, non smoker. No family history of cardiac sudden death, MI, or HTN.

Physical Exam:
- **Vital Signs:** 130/72, HR 85 (regular), RR 18, T 97.8°F, BMI 30
- **Physical Exam:** HEENT – Thyroid wnl   Abd – nontender
  CV – PMI nondisplaced, RRR, no murmurs

Diagnostic/Lab Findings:

12 lead ECG	Normal sinus rhythm
CBC	WNL, no indications of anemia
TSH	WNL, no indications of hyperthyroidism
BMP	WNL, no indications of elec. abnl or ↓gluc
TTEcho	WNL, no structural defects
CXR	WNL, no pul. edema, COPD, Pneumonia

With negative first-line work-up what is the next step?
- Intermittent ambul. ECG monit. w/ loop monitor or event recorder.
- 30-day monitoring results: Self-resolving episodes of HR 130 to 140 coinciding with palpitations and presyncope.

Final Diagnosis: Paroxysmal Atrial Fibrillation (PAF)

- ☑ Episodes last 7 days or less
- ☑ Self-terminating
- ☑ Also classified as lone AF → lacks the classical risk factors for AF.

Take Home Point: Identify and treat other risk factors for AF – in this patient the **EtOH usage** and **"snoring, daytime sleepiness"** should be further explored for alcohol abuse and obstructive sleep apnea.

Pearls: Main Causes of Atrial Fibrillation "I SMART CHAP"

I	**I**nflammatory (Pericarditis/Myocarditis)
S	**S**urgery (Post coronary artery bypass surgery)
M	**M**edications
A	**A**SCHD (Atherosclerotic coronary disease)
R	**R**heumatic valvular disease
T	**T**hyroid
C	**C**ongenital heart disease
H	**H**ypertension
A	**A**lcohol
P	**P**ulmonary Disease* – PE/O2 ↓

*Pearl: Other causes of Atrial fibrillation include:
- Cardiomyopathy, Sepsis, MI, Cocaine, Theophylline, Anemia, *PE, *Hypoxemia [ex. Asthma, COPD], Idiopathic ("Lone AF")

Take Home Point: In addition to managing the rate control of AFIB patients with medical or catheter ablative therapy – daily aspirin vs. anticoagulation is important to **prevent stroke**. The **CHA$_2$DS$_2$-VASc tool** estimates risk of ischemic stroke and guides decisions.

Atrial Fibrillation Stroke Risk: CHADS$_2$VAS$_c$ Score		
	Condition	Points
C	**C**ongestive heart failure	1
H	**H**ypertension	1
A$_2$	**A**ge>75	2
D	**D**iabetes mellitus	1
S$_2$	Prior **S**troke or TIA	2
V	**V**ascular disease (PAD, MI, Aortic plaque)	1
A	**A**ge 65-74	1
S$_c$	**S**ex cat. (female gender w/ other RF)	1

Aspirin [81-325 mg] daily indicated.	Score of **0-1**
Warfarin* indicated for patient.	Score **2 or higher**

*ASA + Plavix acceptable in patients in which Warfarin contraindicated.

Chapter 2: Pulmonary

A. PULM Clinical Pearls

A-a gradient: Simplified Calculation and Pearls

Step 1:	Add PCO_2 + PAO_2 = _____ * *Answer should be approx. 130 (range of 125 – 135)
Step 2:	If **no** A-a gradient is found, it is pure Hypoventilation.
Step 3:	If A-a gradient is more than 135, then patient had to be on O_2

Formulas	Step 1: P_AO_2 = (F_iO_2(Patm -Pwater)) * - (P_aCO_2 / 0.8)
	        = 150 mmHg - (P_aCO_2 / 0.8)
	Step 2: A-a gradient = P_AO_2 - P_aO_2
	        = 150 mmHg - (P_aCO_2 / 0.8) - P_aO_2
	* Hint: At sea level, on room air, the (Patm - Pwater)FiO2 is 150 mmHg; you don't need to calculate this every time.

Allergic Bronchopulmonary Aspergillosis: Clinical Picture

- IgE serum >1000
- Asthma, Pulmonary infiltrates
- Radiology the *glove sign* → "Central Bronchiectasis"
- Precipitating antibodies to Aspergillus antigen
- Scratch test is positive to Aspergillus antigen

Asthma: Samter's Triad

- **Samter's triad** is a medical condition consisting of:
    1. **Asthma**
    2. **Aspirin sensitivity**
    3. **Nasal polyps**

Asthma: Mimickers and Exacerbators

Asthma Exacerbators	Asthma Mimickers	
- Allergic - ASA sensitivity - Chronic sinusitis - Cold - Exertion - Occupation - Stress	- COPD - Foreign body - Sarcoid - Churg-Strauss syndrome - Allergic bronchopulmonary aspergillosis - Cardiac [Mitral stenosis, Cardiomyopathy] - Hypersensitivity pneumonitis (also called Extrinsic allergic alveolitis) - Pulmonary infiltrates and Eosinophilia [Helminths] – PIE syndromes, Löffler's	- GERD - Lung Cancer - Vocal cord paralysis - Eosinophilic pneu.

Cheyne Stokes Respirations: Causes

- Congestive heart failure
- Medications (commonly narcotics)
- Increased ↑ intracranial pressure (CVA, ICH, tumor)

Hemoptysis Pearls
- Massive Hemoptysis criteria is a bleed that is **> 200cc over 24hr**
- Note: **600cc** fills a kidney shaped emesis basin

Test	Diagnostic Clues
WBC	- Shifts – URI, Pneumonia
PT/PTT/INR	- Increased anticoagulant use, coag dis.
D-dimer	- Pulmonary embolism
Sputum gram stain, culture, acid-fast bacillus smear,	- Pneumonia, lung abscess, - TB, mycobacterial infx
HIV	- Increased risk TB, Kaposi sarcoma
ESR	- Infx, Wegener, SLE, sarcoid, Goodpasture syndrome, Neoplasia

Loeffler's Syndrome
- Eosinophilic pneumonia
- Helminthic infection (*A. Lumbricoides*, *N. Americanus*)

Pneumonia, Associated with Special Conditions

Special Conditions	Associated - Pneumonia
Alcoholism	- Klebsiella and aspiration
Dementia	- Aspiration
Nursing home	- Klebsiella, anaerobes
Smoker - COPD	- Haemophilus influenzae and Moraxella
On Steroids or Chemo	- Candida, Staph, PCP, Aspergillus, CMV, HSV, Nocardia
Others	- Pets, hobbies, IVDA, travel

Pulmonary Embolism, Pearls

Classic Symptoms	- Dyspnea - Tachypnea - Pleuritic chest pain - Tachycardia - Hypotension - Hemoptysis (if lung infarct has occurred)
Work-Up	☑ CT scan, pulmonary angiogram ☑ ECG ☑ V/Q scan ☑ X-Ray ☑ D-Dimer
Classic Findings	**Hampton's Hump** – pleural infiltrates corresponding to areas of pulmonary infarction. **Westermark Sign** – CXR reveals dilation of the pulmonary artery proximal to the clot and collapse of the vessels distal to the clot.

Pulmonary Embolism: *Wells Clinical Score for DVT*

Active cancer (treatment ongoing, or within 6 m or palliative)	+1
Paralysis or recent plaster immobilization of the LE	+1
Recently bedridden for >3 d or major surgery <4 wk	+1
Localized tenderness along the distribution of the DV system	+1
Entire leg swelling	+1
Calf swelling >3 cm compared with the asymptomatic leg	+1
Pitting edema (greater in the symptomatic leg)	+1
Previous DVT documented	+1
Collateral superficial veins (nonvaricose)	+1
Alternative diagnosis (as likely or greater than that of DVT)	-2
High probability	>3
Moderate probability	1 or 2
Low probability	≤ 0

*Adapted from JAMA. 1998 Apr 8;279(14):1094-9. Anand SS, Wells PS, Hunt D, Brill-Edwards P, Cook D, Ginsberg JS. Does this patient have deep vein thrombosis?. JAMA. Apr 8 1998;279(14):1094-9.

Pulmonary Manifestations of Lupus

- Myopathy (Shrinking lung syndrome)   - PE (Hypercoagulable state)
- Pulmonary Hypertension   -Serositis / Pleuritis   -Vasculitis / Hemorrhage

Solitary Pulmonary Nodule (SPN)

- **SPN:** an intraparenchymal lung lesion that is < 3 cm in diameter and is not associated with atelectasis or adenopathy.
- **Lung lesions** > 3 cm in size are defined as lung masses.

Character	Radiologic - Benign	Radiologic - MALIGNANT
• Size	• <5 mm	○ > 10 mm
• Border	• Smooth	○ Irregular, Spiculated
• Density	• Dense, solid	○ "Ground glass"
• Calcification	• Homog, Popcorn-like	○ Eccentric or non-calc.
• Doubling time	• <One month, >1year	○ One month to one year

Pearls: Diagnostic Clues for MALIGNANT SPN

- Exposure to Asbestos, tuberculosis, travel, smoker, age (> 50), male, previous malignancy.
- **Likely Etiology** → Adenocarcinoma>SCC>Metastasis>NSCC>Small Cell CA

Pearls: Diagnostic Clues for Benign SPN

- Granulomas (80% benign)
- Calcification and no growth in 2 years.
- **Likely Etiology** → Granuloma>Hamartoma> Aspergillosis> Coccioidomycosis>Cryptococcosis>Histoplasmosis> Tuberculosis

Pearl: Cavitary Lesions on Chest X-Ray

Top 4 causes of Cavitary Lesion on Chest X-ray

1. Vasculitis
2. Infection
3. Malignancy
4. Pulmonary embolism

- **Thick** walled = **Tuberculosis**   -----vs-----   Thin walled = *Coccidiomycosis*

B. PULM Differential Diagnosis

Hemoptysis: DDx

- Abscess
- Arteriovenous malformation
- Aspergilloma
- Bronchiectasis
- Bronchitis
- Congestive heart failure
- Coagulopathy
- Cocaine
- Cystic fibrosis
- Foreign body
- Mitral stenosis
- Lupus pneumonitis
- Pulmonary embolism
- Primary pulmonary hemosiderosis
- Pulmonary-renal syndromes
- Septic emboli
- Tuberculosis / Mycobacterium avium-intracellulare/Fungal
- Vasculitis
- Neoplasm

Rare causes of hemoptysis
- Lymphangioleiomyomatosis
- Yellow nail syndrome
- Pulmonary endometriosis

Multiple Pulmonary Infiltrates: DDx

Infections	Inflammatory or Allergic	Neoplastic
<ul><li>Bacterial Pneumonia</li><li>Fungi<ul><li>Endemic mycoses</li><li>Aspergillosis</li></ul></li><li>Eosinophilic pneumonia<ul><li>Ascariasis</li><li>Paragonimus</li><li>Strongyloides</li></ul></li><li>Parasites</li><li>Pneumocystis jiroveci</li><li>Tuberculosis</li><li>Tropical eosinophilic pneumonia</li><li>Filariasis</li></ul>	<ul><li>Allergic bronchopulmonary aspergillosis</li><li>Bronchiolitis obliterans organizing pneumonia</li><li>Churg-Strauss syndrome</li><li>Drug reactions</li><li>Sarcoidosis</li></ul>	<ul><li>Bronchoalveolar cell carcinoma</li><li>Lymphoma</li></ul>

Pneumonia Atypical Causes: DDx

- Chlamydia
- Fungal
- Legionella
- Pneumocystis Jiroveci (i.e. HIV related)
- Tuberculosis
- Viral
- Mycoplasma

Unresolving Pneumonia: DDx

- Adult respiratory distress syndrome (ARDS)
- Bronchiolitis obliterans with organizing pneumonia
- Cancer
- Congestive heart failure
- Hemorrhage
- Empyema
- Pulmonary embolism
- Radiation
- Sarcoid
- Systemic lupus erythematosus
- Vasculitis
- Wrong drug and/or Wrong organism

Pulmonary Hypertension: DDx

- Cocaine
- COPD
- Diet medications (ex. Fenfluramine, Phentermine)
- Hepatopulmonary (syndr.)
- Left heart failure
- Left-to-right shunt
- Obesity
- Primary pulmonary htn.
- Pickwickian syndrome (Obesity hypoventilation syndrome)
- Pulmonary embolism
- Rheumatoid arthritis
- Scleroderma
- Sleep apnea
- Vasculitis
- Interstitial lung disease
- Chronic venous thromboembolic disease in pulmonary circulation

Sinobronchial Syndromes: DDx

- Allergic bronchopulmonary aspergillosis
- Allergic fungal sinusitis
- Amyloid
- Ascariasis
- Arteriovenous malformations
- Churg-Strauss syndrome
- Cystic fibrosis
- Common variable immunodeficiency (CVID)
- IgA deficiency
- IgG subclass deficiency
- Immotile cilia syndrome
- Invasive aspergillus
- Wegener's granulomatosis

Solitary Pulmonary Nodule (SPN): DDx

- Bronchiolitis obliterans organizing pneumonia / Cryptogenic organizing pneumonia
- Coccidioidomycosis
- Cyst
- Dirofilaria immitis (the dog heart worm, perfect coin lesion appearance)
- Echinococcus
- Fibroma
- Hamartoma
- Histoplasmosis
- Lipoma
- Pneumocystis Jiroveci pneumonia
- Pulmonary embolism
- Rheumatoid arthritis
- Sarcoid
- Tuberculosis
- Wegener's granulomatosis

C. PULM Mnemonics

Asthma Exacerbation

D	Drugs*	
I	Infections (URTI/LRTI)	
P	Pollutants (cig smoke, perfume)	
L	Labile mood (emotion)	
O	esOphageal reflux (GERD)	
M	Mites (dust allergens)	
A	Activity (exercise)	
T	Temperature (cold)	

DIPLOMAT
Precipitating factors for **asthma exacerbation**.

*Drugs:
- Beta blockers
- Aspirin
- NSAIDs

Chronic Cough

TB	Tuberculosis
C	Cigarette smoking (Lung CA)
C	Congestive heart failure
O	Obstructive lung dz (asthma, COPD)
U	Upper airway (sinus., pharyn., otitis)
G	GERD
H	Hypertension medications (ACEi)

TB CCOUGH
Common causes of **chronic cough**.

Hemoptysis

C	Congestive Heart Failure
A	Airway disease (Bronchitis)
V	Vasculitis/Vascular malfor.
I	Infection (TB, Abscess, Pneumo, Fungal Mycetoma)
T	Trauma (foreign body)
A	Anticoagulation
T	Tumor (malign, KS, carcinoid)
E	Embolism (PE, septic emboli)
S	Stomach (GI Bleed)

CAVITATES
Common causes of **hemoptysis**.

Others: Lupus pneumonitis, Idiopathic pulmonary hemosiderosis, Goodpastures, Wegeners.

Rare causes of hemoptysis
- Lymphangioleiomyomatosis
- Yellow nail syndrome
- Pulmonary endometriosis

Lung Cancer

B	Bone
L	Liver
A	Adrenals
B	Brain

BLAB
Common Sites for **Metastasis from 1° Lung Cancer.**

Pulmonary Edema

N	Near Drowning	**NOT CARDIAC**
O	O2 Therapy / Post Intubation	Non-cardiac causes
T	Trauma/Transfusion (TRALI)	of **pulmonary edema.**
C	CNS related (neurogenic pulm edema)	
A	Allergic Alveolitis	
R	Renal Failure	
D	Drugs	
I	Iatrogenic (IVF overload), Inhalation	
A	high Altitude pulm edema / ARDS	
C	Contusion	

Pulmonary Fibrosis

SCAR: Causes of Pulmonary Fibrosis in Upper and Lower Lobes.

UPPER		LOWER	
S	Silicosis/ Sarcoidosis	S	Systemic sclerosis
C	Coal worker pneumonconiosis	C	Cyptogenic fibrosing alveolitis
A	Ankylosing spondylitis / Allergic alveolitis / ABPA / (also TB)	A	Asbetosis
R	Radiation (not shielded)	R	Rheumatoid arthritis / Radiation (not shielded)

Shortness of Breath/Dyspnea

C	Congestion (CHF)	***CD-SPIES***
D	Drugs (Narcs, Benzos)	Differential
S	Spasms (Broncospasm, COPD, Asthma)	diagnosis of
P	Pneumothorax	**SOB/Dyspnea.**
I	Infection (Pneumonia)	
E	Embolism (Pulmonary embolism)	
S	Secretions (Mucous plugging)	

Tuberculosis

R	Rifampin	**RIPE**
I	Isoniazid	**Multidrug** treatment
P	Pyrazinamide	regimen for **tuberculosis.**
E	Ethambutol	

D. PULM Clinical Case #1

History:
- 2 month history of persistent pneumonia in a 30-year old male with negative preliminary work-up (below), s/p treatment with multiple antibiotics including macrolides and doxycycline.

Diagnostic/Lab Findings:

Diagnostic/Lab Findings:	
Multiple sputum gram stain	Neg
AFB	Neg
HIV testing	Neg
Cytology	Neg

Differential Diagnosis of Persistent and Unresolving Pneumonia:
Unresolving Pneumonia: DDx
- Acute respiratory distress syndrome
- Bronchiolitis obliterans organizing pneumonia / Cryptogenic organizing pneumonia
- Cancer
- Empyema
- Hemorrhage
- Pulmonary Embolism
- Radiation
- Sarcoid
- Systemic lupus erythematosus
- Vasculitis
- Wrong Organism (BUG) and/or Wrong Drug

Clinical Pearl:
- **History-Taking Skills:** Must take thorough history including **occupation, travel, and pet exposure details**. The historical clue that lead to a diagnosis in this case was **travel** to the **San Joaquin Valley**.

Final Diagnosis:
- **Coccidioidomycosis.**

Pearls: Coccidioidomycosis
- **Coccidioidomycosis** (San Joaquin fever, valley fever) is an infection, usually of the lungs, caused by the fungus ***Coccidioides immitis***. It is endemic in *California, New Mexico, Arizona, and Nevada*.
- Source:http://www.cdc.gov/nczved/divisions/dfbmd/diseases/coccidioidomycosis/

D. PULM Clinical Case #2

History:
- 40-year old female with history of asthma for 2 years. Also diagnosed with allergic rhinitis 1 year ago. Presents to ED with a severe asthma attack, purpuric type rash over her lower extremities, and a right foot drop. Pt denies history of smoking.

Laboratory Findings:

ACE level	Normal
CBC	Mildly elevated WBC but marked eosinophilia
D-Dimer	Negative
Echocardiogram	Normal
Sputum Cultures	Negative
Chest X-Ray	Bilateral pulmonary infiltrates, No hilar adenopathy

Clinical Pearl: Asthma Exacerbators and Mimickers

Asthma Exacerbators	Asthma Mimickers
<ul><li>Allergens (perfume, smoke, dust, pollen, pets)</li><li>ASA sensitivity</li><li>Chronic sinusitis</li><li>Cold</li><li>Exertion</li><li>Occupation</li><li>Stress / Panic Atck.</li></ul>	- COPD - GERD - Foreign body - Lung Cancer - Sarcoid - Vocal cord paralysis - Churg-Strauss syndrome - Eosinophilic pneu. - Allergic bronchopulmonary aspergillosis - Cardiac [Mitral stenosis, Cardiomyopathy] - Hypersensitivity pneumonitis (also called Extrinsic allergic alveolitis) - Pulmonary infiltrates and Eosinophilia [Helminths] – PIE syndromes, Löffler's

Final Diagnosis:
- Churg-Strauss Syndrome

Clinical Summary – Churg-Strauss Syndrome
- **Churg-Strauss Syndrome** [aka, Allergic Granulomatosis] is a rare systemic vasculitis consisting of asthma, eosinophilia, fever, and accompanying vasculitis of various organ systems.
- **Take home point: "All the wheezes is not always asthma."** While asthma and COPD are common causes, be aware of the mimickers.
- **Take home point:** Asthma triads, Asthma & two clinical features:

Churg-Strauss	Asthma + Eosinophilia + Granulomatous vasculitis
Samter's triad	Asthma + Aspirin sensitivity + Nasal polyps
ABPA	Asthma + Pulm infiltrates+ allergic to Aspergillus

Chapter 3: Gastroenterology

A. GI Clinical Pearls

Charcot Triad

1. Fever/Chills
2. Jaundice
3. RUQ pain

Triad supports diagnosis of:
Cholangitis =infection of common bile duct

Constipation: Causes, Work-up, Prevention/Treatment

Endocrine	<ul><li>Diabetes mellitus</li><li>Hypercalcemia</li><li>Hypomagnesium</li><li>Hypothyroidism</li></ul>	<ul><li>Pregnancy</li><li>Primary Hyperparathyroidism</li><li>Renal Failure</li></ul>
Meds	<ul><li>Alum + Ca2+ antacid</li><li>Anesthetic</li><li>Anticholinergic</li><li>Antidepressant</li><li>Antiepileptic Drugs</li></ul>	<ul><li>Bismol</li><li>Iron Sulfate</li><li>Opiates</li><li>Resins/Bile acid Resins</li><li>Vinca alkaloids</li></ul>
Neuro	<ul><li>Cauda Equina Syndrm.</li><li>Hirschsprung's disease</li></ul>	<ul><li>Myotonic dysfunction</li><li>Spinal cord injury</li></ul>
Structural	<ul><li>Adhesions</li><li>Diverticula</li><li>Ulcerative colitis</li><li>Crohn's disease</li></ul>	<ul><li>Tumors</li><li>Rectocele</li><li>Systemic Sclerosis / Scleroderma</li></ul>
Work-Up	**Labs:** ☑ Hemoccult ☑ Calcium, Electrolytes ☑ TSH ☑ Urine drug screen	**Diagnostics:** ☑ Anorectal manometry ☑ Colonoscopy ☑ CT ☑ Defecography
Prevention -----and----- Treatment	<ul><li>**Bulk Laxatives:** Psyllium and Water (20-30g daily).</li><li>**Stool Softner:** Docusate Sodium (Colace).</li><li>**Osmotic:** Magnesium Citrate, Magnesium Hydroxide (Milk of Magnesia), Polyethylene glycol (Miralax)</li><li>**Stimulant:** Bisacodyl (Dulcolax), Senna (Senokot)</li></ul>	

Diarrhea: Likely Sources in HIV Patients

Take Home Message: HIV pts w/ diarrhea 1st work-up the *usual suspects* -- then test for the following likely sources:

- Cryptosporidium
- Cytomegalovirus
- Isospora belli
- Lymphoma
- Microsporidium
- Mycobacterium avium-intracellulare

Diarrhea: Types, Pearls, Management

Osmotic
- Lactose deficiency
- Laxative

Malabsorption
- AIDS
- Bacterial overgrowth
- C. *difficile*   Celiac sprue
- Giardia   Lymphoma
- IBD   Pancreatitis
- Parasites   Whipple's dz

Endocrine
- Addison's disease
- Diabetes mellitus
- Hyperthyroidism

Secretory
- Carcinoid
- Hyperthyroid
- Villous adenoma
- VIPoma
- Zollinger-Ellison syndrome

Other
- Collagenous colitis

Etiology PEARLS:
- **Fever & Blood in stool** → Shigella, Campylobacter, Salmonella (may be without blood), Escherichia coli 0157:H7.
- **Fever and NO Blood** → rotavirus, Norwalk, enterotoxic E. coli, food poisoning (Staph aureus, Clostridium perfringens), Vibrio cholera.

Management Pearls:
- AVOID **Loperamide** in patients with **fever** or **blood** in stool.
- Antibiotics indicated – Pt fever + bloody stools (quinolones), traveler's diarrhea, C. difficile infection (metronidazole)

Discriminant Function [Alcoholic Hepatitis]	
Formula	4.6 **x** [Patient's Prothrombin Time – Control Prothrombin Time] **+** Serum Total Bilirubin _____ *
Pearl	_____*>**32 points** indicates poor prognosis and patient may benefit from glucocorticoid therapy.
Treatment	☑ Treatment: **Methylprednisolone** 32 mg q day for 28 days

Food Poisoning Syndrome Toxins	
Likely Agents Producing Toxins	**Likely Source**
Bacillus cereus	Chinese rice
Staphylococcus aureus	Ham/Mayonnaise
Clostridium-induced botulism	Canned foods, Jams
Vibrio Parahaemolyticus	Raw seafood
Trichinella Spiralis	Raw meat

GI Bleed: LOWER GI Bleed Pearls	
Top 2 Causes of Bloody Diarrhea	• Diverticulosis • Arteriovenous malformation • Others: IBD (Crohn's, UC), Infection **NOTE:** Must rule-out ischemia
HPI Pertinent +'s	+ NSAIDS / Aspirin use / Alcohol use + Warfarin use + Melena vs. hematochezia
PMHx Pertinent +'s	+ Cancer + Recent Surgeries + Cardiac risk factors [Aortic graft, A-Fib] + Change in stool [Pus, blood, mucous]

GI Bleed: UPPER HPI Pertinent Positives		
+ Coumadin	+ Dysphagia	+Steroids + Pills
+ Dysphagia	+ Syncope	+Vomiting prior
+ NSAIDs/ASA/EtOH	+ Nose bleeds	+ Weight loss

Hepatitis B Markers	
Serum Markers	**Interpretation**
Hepatitis B surface Antigen (**HbsAg**)	• **Active** infection
Hepatitis B surface Antibody (**HbsAb**)	• **Past** infection ----or----- • **Vaccination** against hepatitis B
Hepatitis Be Antigen (**HbeAg**)	• Active **replication** of the virus
Anti Hepatitis B core IgM Antibody (**Anti-HBc IgM**)	• **Acute** infection
Anti Hepatitis B core IgG Antibody (**Anti-HBc IgG**)	• **Chronic** infection

- **PEARL:** Hepatitis B **Chronic** AND **Carrier** patients have **HbsAg** and **Anti-HBc IgG**. To distinguish <u>check LFTs</u>.
- **Carrier** = normal LFTs   vs   **Chronic** = ↑ **LFTs**.
- **PEARL:** Hepatitis D requires coinfection with **Hepatitis B**. Most dangerous coinfx w/ **Anti-HBc IgG** (acute super infx, fulminant hep.)

Hepatitis C: Extra-Hepatic Manifestations	
• Cryoglobulinemia • Idiopathic thrombo. purpura • Lichen Planus • Lymphoma, MGUS, Mult. myelo. • Membranoproliferative glomerular nephritis (MPGN)	• Mooren's corneal ulcers • Porphyria cutanea tarda • Sialoadenitis • Sjögren's syndrome • Thyroiditis • Vitiligo

Liver, LFTs Patterns

Take Home Message: Usually the AST and ALT are not > 300-400 in pure alcohol-related liver disease. If transaminases are >400 the clinician must look for other causes.

LFTs	Interpretation
Hepatocellular	AST/ALT **3:1** → **Alcohol-related** Hepatitis
AST	Other than liver disease, AST elevations can be found in heart, bone, and muscle may be due to **hemolysis**.
Alk Phos↑	**Cholestatic DDx (Alk Phos↑ elevated out of proportion to the transaminases):** • AIDS cholangiopathy • Cancer • Mets • Drugs • Granulomatous (ex. Sarcoid, Tuberculosis) • Primary biliary cirrhosis • Primary sclerosing cholangitis • Gallstone calculi • Sepsis • Syphilis
Transaminases ↑ AST and ALT	**Hepatocellular DDx (transaminases↑ elevated out of proportion to Alk Phos):** • Alcohol Toxins • Autoimmune hepatitis • Celiac Sprue • Drugs (Statins & Tylenol) • Non-alcoholic fatty liver • Musc. injury, CHF, Malig • Viral infection -(Hep ABCDE) - plus EBV, CMV, HSV • Iron disorder -Wilsons, alpha1-antitrypsin deficiency, hemochromatosis
Low albumin	• Acute or chronic **inflammation** (most common), **urinary loss**, severe **malnutrition** or **liver** disease. • **GI loss** (colitis or uncommon small bowel disease)
↑ Serum Ammonia Level	• **PEARL: not** necessarily elevated in pts w/ hepatic encephalopathy. • Useful test in pt w/**new onset altered mental status**.
↑ Isolated Uncong Bilirubin	• Consider **Gilbert syndrome*** or **hemolysis**. *episodes of hyperbilirubinemia exacerbated by stress (due to a defic. in bilirubin-UGT enzyme)
↑ Isolated γ-glutamyltransf.	• May be induced by **alcohol** and **aromatic meds**. • No actual liver disease

Liver, Toxic Over-the-Counter Supplements

- Chaparral
- Comfrey
- Germander
- Hydroxycut
- Jin Bu Huang
- Kava Kava
- Lipokinetics
- Lots of Co Enzyme Q
- Ma Huang
- Pennyroyal
- Sassafras
- Shark cartilage
- Weight loss products (select types)

Nausea and Vomiting, Intractable: Workup

- ☑ Amylase and Lipase
- ☑ CT abdomen
- ☑ Esophagogastroduodenoscopy
- ☑ Gastric emptying scan
- ☑ Liver function tests
- ☑ MRI
- ☑ RUQ - Ultrasound
- ☑ Small bowel follow through
- ☑ Upper GI series

Pancreatitis

Five things to look for in patients w/ severe pancreatitis		
1.	↓ Decrease PO_2	ARDS
2.	↓ Hemoglobin	Hemorrhagic pancreatitis
3.	↑ Increased Glucose	Pancreatic insufficiency
4.	↓ Decreased Ca^{2+}	Saponification
5.	↑ Increased BUN/Cr	Third spacing + Fluid sequestration

Other causes ↑ Amylase besides pancreatitis:

- Cholecystitis
- DKA
- Ectopic pregnancy
- Lung – Ovary - Salivary Cancer
- Perforation - PUD
- Renal Failure
- Salpingitis
- Intestinal infarct

PUD: Endoscopic Characteristics of PUD Upper-GI Bleeding and Rates of Rebleeding

Endoscopic Characteristic	Management
55% A Actively bleeding visible vessel	☑ Need NPO ☑ Monitor in hospital
43% N Nonbleeding visible vessel	☑ Need NPO ☑ Monitor in hospital
22% A Adherent clot	☑ PO ☑ Rx PPI + discharge home
10% F Flat pigmented lesion	☑ PO ☑ Rx PPI + discharge home
5% C Clean ulcer base	☑ PO ☑ Rx PPI + discharge home

B. GI Differential Diagnosis

Ascites, DDX: (based on SAAG*)

High Gradient (>1.1 g/dL = Portal hypertension)	Low Gradient (<1.1 g/dL = Nonportal hypertension)
- Cirrhosis	o Peritoneal carcinomatosis
- Portal vein thrombosis	o Tuberculous peritonitis
- Budd-Chiari syndrome	o Pancreatic ascites
- Congestive heart failure	o Bowel obstruction
- Constrictive pericarditis	o Bowel infarction
- Myxedema	o Serositis (e.g. in Lupus)
- Massive hepatic metastases	o Nephrotic syndrome
SAAG - Serum Ascites Albumin Gradient **SAAG** Formula = Serum albumin – Ascitic albumin = _____ g/dL	

Cramping Disorders: DDx

- Alcohol
- Cirrhosis
- Carnitine palmityl transferase deficiency
- Dialysis
- Glycogen storage diseases including:
  - o McArdle's disease (i.e., myophosphorylase deficiency)
  - o Phosphofructokinase (PFK) deficiency
- Heat
- Hypoglycemia ($\downarrow$ Blood Glucose)
- Hypocalcemia ($\downarrow$ Ca^{2+})
- Hypomagnesemia ($\downarrow$ Mg^{2+})
- Hyponatremia ($\downarrow$ Na^+, $\downarrow$ Salt)
- Medications may cause cramps (ex. Lithium, Nifedipine, Statins, Terbutaline)
- Pregnancy
- Respiratory alkalosis
- Thyroid

Gastric Folds, Thickened: DDx

- Amyloidosis
- Lymphoma
- Malignancy
- Ménétrier's disease
- Syphilis
- Zollinger-Ellison syndrome

Gastric Lesions, Multiple: DDx

- Ischemia
- Lymphoma
- Zollinger-Ellison syndrome

Upper Gastrointestinal Bleed: DDx

- Cancer
- Dieulafoy's
- Esophagitis
- Gastritis
- Mallory Weiss tear
- Peptic ulcer disease
- Varices

Lower Gastrointestinal Bleed: DDx

Lower GI Bleed with PAIN	Lower GI Bleed Painless	Other DDx
- Ischemia - Inflammatory bowel disease	o AVM o Diverticulosis O Hemorrhoids o Rapid upper GI bleed	- Henoch Schönlein Purpura - Cancer

Nausea and Vomiting: DDx

- Addison's disease
- CNS – MRI, MS, tumor
- Gall bladder
- Gastroparesis
- GERD
- Hypercalcemia ↑Ca^{2+}
- Hypernatremia ↑Na^+
- Hyponatremia ↓Na^+
- Medications
- Mesenteric ischemia
- Pancreatitis
- Peptic ulcer disease
- Pregnancy
- Scleroderma
- Superior mesenteric artery (SMA) syndrome (***Pearl: auscultate for epigastric bruit**)
- Uremia

Pancreatitis: DDx

General causes:
- Alcoholism
- Calcium (↑Ca^{2+})
- Cancer
- Hereditary pancreatitis
- Hypertriglyceridemia
- Mumps
- Pancreatic divisum
- Post- endoscopic retrograde cholangiopancreatography (ERCP)
- Renal calculi
- Trauma

Other potential causes:
- HIV, Coxsackie B, EBV, Autoimmune pancreatitis

Rare:
o Trinidadian Scorpion Envenomation

*Pearl: In the United States over 90% of pancreatitis is caused by either **gallstones** or **alcoholism**. The 3rd leading cause of pancreatitis in the US is **medications**.

Pancreatitis caused by Medications: DDx

- 6-Mercaptopurine
- Depakote
- Diuretics
- Estrogen
- Imuran
- Pentamidine, ddl, ddc
- Oral hypoglycemics
- Tetracycline, Sulfa
- ☑ ***Pearl: Always check the medications list.** Medications are the 3rd leading cause of pancreatitis in US.

Portal Vein Thrombosis, Risk Factors: DDX

Liver Specific	Liver Cirrhosis	
Local Factors	**Abdominal infection or sepsis** - Appendicitis - Cholangitis - Cholecystitis - Diverticulititis - Pancreatitis - Abdominal abscess	**Abdominal surgery** - Especially for infection - IBD - Adominal malignancy - Abdominal trauma
Hypercoaguable States, Thrombophilia	- APLS - ATIII defic. - Bechet - Cancer - Factor V Leiden - HIT - IBD - JAK Z poly.	- Monoclonal gammopathy - Myeloproliferative disorder - OCP use - PT20210 mutation - Pregnancy - PNH - Pro C, S deficiency

Splenomegaly, Massive: DDX

Infectious	- **Kala azar (visceral leishmaniasis).** **Pearl:** H&P (+'s) Sub-saharan Africa, Asia, Mediterranean, Latin America. Transmitted by sandflies. Macrophages infected- main reservoirs of infxn in spleen liver and bone marrow. Latency- 2-8 months. Intermittent fevers, pancytopenia, hepatosplenomegaly, elevated IgG. - **Hyperreactive malarial splenomegaly syndrome** **Pearl:** (aka. tropical splenomegaly syndrome), Abnormal immunologic response to malarial infection. Requires pancytopenia and IgM elevation for dx. - Echinococcal cysts (rarely) AIDS with MAC infxn
Infiltrative	- **Gaucher Disease**: Lysosomal glucocerebrosidase deficiency. (↑Ashkenazi Jews). - **Niemann-Pick disease**: Autosomal recessive lysosomal storage disease. HSM + neuro deficits.
Hematologic	- CML - CLL - Hodgkins - Heavy chain disease - Polycythemia vera - POEMS (polyneuropathy, organomegaly, endocrinopathy, M protein, and skin lesions). - Amyloid - Waldenstroms - Multiple myeloma - Myelofibroisis

C. GI Mnemonics

Abdominal Pain: NON-Appendiceal RLQ Pain

A	Abscess	**APPENDICITIS** is **Not** the **Only** cause of RLQ pain.
P	Pancreatitis	
P	Pelvic inflammatory disease	
E	Ectopic pregnancy	**NON-Appendiceal Causes** for **right lower quadrant** abdominal pain
N	Neoplasia	
D	Diverticul-itis/-osis	
I	Intussusception	
C	Cholecystitis	Others: Omental torsion, Endometriosis, Enterolitis, degenerating uterine leiomyoma, colon cancer, mesenteric adenitis
I	Inflammatory Bowel Disease	
T	urinary Tract infection	
I	Irritable Bowel Syndrome	
S	Stones	
Not	**Not** having BMs (constipation)	
Only	**O**varian (torsion, cyst, abscess)	

Cholelithiasis

F	Female	**5 Fs**
F	Fertile	Risk Factors for **Cholelithiasis**.
F	Fat	
F	Forty years old	
F	Family History	

Hepatitis, Causes

A	Acetaminophen or viral Hepatitis **A**	**ABCDE** Main causes of **Hepatitis**
B	Hepatitis **B**	
C	Hepatitis **C**	
D	Drugs*	* NSAIDs, OCPs, statins, sulfa drg., erythromycin, tetracyclines.
E	Extras (Wilson's disease, Autoimmune hepatitis, Hemochromatosis, Alpha-1-antitrypsin deficiency)	

Hepatitis C

2%	Risk of monogamous sexual transmission	**5 Deuces of Hepatitis C**
2%	of the population has it = 3.9-4 mill.	5 features of **Hepatitis C** at 2% occurrence
2%	Needle stick risk	
2%	Risk of vertical transmission	
2%	Chance of hepatoma per year/after cirrh.	

Hepatitis, Drug-Induced

F	Fever
A	Arthralgia
R	Rash
E	Eosinophilia

FARE
Clinical picture of **Drug-Induced Hepatitis** and can also be used for **Acute Interstitial Nephritis**.

Hepatocellular Cancer

W	**W**ilson's disease
AT	**A**lpha-1-anti**T**rypsin
C	**C**arcinogens*
H	**H**emochromatosis
A	**A**lcoholic cirrhosis
B	Hepatitis **B**
C	Hepatitis **C**

WATCH for ABC
Major Risk Factors of **Hepatocellular Cancer**

*aflatoxin B1, polyvinyl chloride

Ileus

P	**P**ostoperative
P	**P**eritonitis
P	**P**otassium is low
P	**P**elvic and spinal fractures
P	**P**arturition

5 Ps
Causes of **Paralytic Ileus**.

Meckel's diverticulum

2	**2** times more common in Males
2	Occurs within **2** feet from ileocecal valve
2	**2** types of ectopic tissue (gastric and pancreatic)
2	Found in **2%** of population
2	Most complications occur before age **2** years old.

Rule of 2s
Pattern of "2s" in **Meckel's diverticulum**

Needle Stick Accidents

HIV	0.3%
Hep C	3.0%
Hep B	30.0%

3s of Needle Stick Accidents
Hepatitis B, C, and HIV: If you are **stuck with a needle** that has fluids from a patient HIV, Hep C, or Hep B, then approx risk of contraction is listed.

Non-Alcoholic Fatty Liver Disease

D	**D**yslipidemia and **D**iabetes mellitus
R	Insulin **R**esistance
O	**O**besity
P	**P**ressure (i.e. Hypertension)

NAFLD-DROP
Causes for **Non-Alcoholic Fatty Liver Disease**

Pancreatitis

G	Gallstones	***GET SMASHeD***
E	Ethanol, ERCP	Causes of **Acute**
T	Trauma	**Pancreatitis**
S	Steroids	
M	Mumps (viruses)	
A	Autoimmune disorder	
S	Scorpion sting	
H	Hyperlipidemia	
D	Drugs (especially didanosine DDI)	

Vomiting

V	Vestibular disturb. / Vagal -reflex pain	**VOMITTING**
O	Opiates	Differential diagnosis
M	Migraine / Metabolic*	for **Extra-GI Vomiting**.
I	Infection**	*Metabolic Causes:
T	Toxicity (cytotoxic, digitalis toxicity)	DKA, gastroparesis,
T	Too much eTOH	hypercalcemia
I	Increased intracranial pressure	**Infx Causes:
N	Neurogenic, psychogenic	TB, Abscess, PNA,
G	Gestation	Fungal, Viral.

Chapter 4: Renal & Genitourinary

A. Renal Clinical Pearls

Acute Renal Failure: Prerenal, Intrarenal, Post-Renal

Pre-renal	↓ Decreased cardiac output due to CHF ↓ Low albumin states (i.e. Nephrotic syndrome) ↓ Low effective circulating volume • Cirrhosis Hepatorenal / Anemia / Dehydration
Intrarenal	• Acute interstitial nephritis (Fever, Rash, Eosinophilia) • Acute tubular necrosis (↓Blood pressure, Sepsis) • Contrast / Glomerulonephritis / Pigment / Vascular
Post-Renal	• Benign prostatic hyperplasia • Bladder stone • Neurogenic bladder • Ovarian or cervical cancer • Retroperitoneal fibrosis • Retroperitoneal lymphadenopathy

Chronic Kidney Disease Stages

Chronic Kidney Disease Stage	Glomerular Filtration Rate
Stage 1	< 90
Stage 2	60-89
Stage 3	30-59
Stage 4	15-29
Stage 5	< 15

Chronic Renal Failure, Top 5 Causes

- Hypertension
- Glomerulonephritis
- Diabetes mellitus
- Obstructive uropathy
- Autosomal dominant polycystic kidney disease [ADPKD]

Chronic Renal Failure Patients, "Golden Rules"

#1	Use *renal dosing* for **all** medications.
#2	Avoid / limit use of *contrast* agents. • Including MRI with gadolinium • Patients to be especially watchful for: 　○ **Diabetics** w Cr>1.5 　○ **Multiple myeloma** 　○ **Sickle cell disease**
#3	Avoid *nephrotoxic medications.* Examples: • NSAID　　• Chemo • ACE-I　　• Amphotericin B • Gentamycin

Contrast-Induced Nephropathy

Patients w/ ↑ *increased* risk of contrast induced nephropathy:
- Diabetic patients w/ Creatinine > 1.5
- Multiple myeloma patients
- Sickle cell patients

Options to help prevent contrast-induced nephropathy:
- IV fluid hydration is the **best** preventative measure.
- ↓ Osm contrast agents
- Avoid ventriculogram
- Administer Mucomyst

Dialysis: Absolute Indications

- Fluid overload – refractory to diuresis
- Pericarditis
- Refractory acidosis
- Refractory hyperkalemia
- Severe uremic symptoms (neurological signs, seizure, coma)
- Toxins/Toxin Ingestion

Nephrotic Syndromes, Criteria, and Clinical Picture

Clinical Picture	1° Renal Causes	2° Renal Causes

- >3.5 gm in 24 hours
- ↑ Cholesterol
- ↓ Albumin
- Edema
30% Adults is secondary SLE, Diabetes mellitus, Amyloid.
70% is primary.
	- Focal Segmental
	- IgA nephropathy
	- Membranoproliferative
	- Membranous
	- Mesangioproliferative
	- Minimal Change
	- Post-infectious
		- Amyloid
		- Diabetes Mellitus
		- Hypothyroid
		- NSAIDs
		- Systemic lupus erythematosus
		- Viral
		  ○ Hepatitis B
		  ○ Hepatitis C
		  ○ HIV Nephropathy

Nephrotic Syndrome, Workup

- ☑ ANA
- ☑ ASO titer
- ☑ C3 and C4
- ☑ HbA1c
- ☑ Hepatitis panel
- ☑ HIV
- ☑ Renal Ultrasound
- ☑ Renal biopsy for diagnosis followed by:
  ○ light microscopy
  ○ electron microscopy
  ○ immunofluorescence
- ☑ Serum protein electrophoresis
- ☑ Urine protein electrophoresis

***Always** check **medication list**

Renal Failure, Bone Diseases

- Osteitis fibrosa cystica
- Osteoporosis and osteomalacia

Renal Artery Stenosis: Clinical Clues
<ul><li><20 years old</li><li>Flash pulmonary edema</li><li>Renal artery bruit auscultated on PE</li><li>Associated with PVD</li></ul> **Caution:** Worsening of renal function with use of Angiotensin-converting enzyme (ACE) inhibitors, other anti-hypertensive meds, and diuretics.

Urinary Retention, Post-Operative	
1st Steps Prior to Urology Consult:	Management
<ul><li>Check medication list*</li><li>Prostate exam</li><li>Urinalysis</li><li>**Do not** check PSA in the acute setting.</li></ul>	☑ I/O cath every 6 hours ☑ Trim-Sulfa, 1 tab PO BID- UTI PPx ☑ 6 wk trial Flomax or Alpha-Blocker ☑ 2 wk f/up check for improvement [can ↑increase or ↓decrease med based on pt status] ☑ Then Cysto and Turp with Urology consult if no resolution

*Check med list:** For agents causing urinary retention.

B. Renal/GU Differential Diagnosis

Glomerular Disease in Patients with HIV / Hep C: DDx

Hepatitis C-related	HIV-related
▪ Fibrillary or Immunotactoid GN ▪ Membranoproliferative GN ▪ Membranous GN ▪ Mixed cryoglobulinemia	○ Collapsing focal segmental glomerulosclerosis

Nephrotic Syndrome: 1° Renal Etiologies vs 2° Causes: DDx

1° Renal Etiologies	2° Renal Etiologies
▪ Focal segmental glomerulosclerosis ▪ Membranoproliferative glomerulonephritis ▪ Membranous glomerulonephritis ▪ Mesangial proliferative glomerulonephritis ▪ Nil change	▪ Amyloid ▪ Diabetes mellitus ▪ NSAIDs ▪ Post-streptococcal glomerulonephritis ▪ Systemic lupus erythematosus ▪ Viral → Hepatitis B and C

Nephrotic Syndrome: Other Causes

- Alport's syndrome
- Bacterial endocarditis
- Chronic pyelonephritis
- Dengue fever
- Fabry's disease
- Heavy metals
  - Arsenic
  - Copper
  - Lead
  - Mercury
- Malaria
- Multiple myeloma
- Polycystic kidney disease
- Renal sarcoid
- Sepsis
- Shigella
- Tuberculosis

Pearl: Other clinical manifestations of Nephrotic Syndrome: ↓Decrease in C3, ↑Lipids, Thrombosis, ↑Increase infection

Renal Cysts: DDx

- AD polycystic kidney disease
- Medullary sponge kidney
- Primary renal sarcoma
- Renal lymphangiomas
- Renal-cell CA w/ cystic changes
- Simple cysts
- Tuberous sclerosis
- Von Hippel-Lindau syndrome

Post-Op Urinary Retention: DDx

- Anesthesia
- Anticholinergics
- Benign prostate hyperplasia
- Decongestants
- Medications
- Neurogenic bladder
- Narcotics

C. Renal/GU Mnemonics

Acute Renal Failure

Corp	**C**onnective tissue disorders SLE, MCTD, HUS, TTP, PAN, Wegener's, GP	**Corporate Vice Presidents Hate Dogs Mating** DDx of acute renal failure
VICE	**V**ascular, **B**ilateral RAS, **E**mboli, **I**schemic, Renal vein thrombosis, **I**nfection – severe pyelo/sepsis	
Presidents	**Pre**renal, **Pre**gnancy, **Pre**-Eclamp.	*NSAID, contrast, Gentamycin, ACEi, Chemo, Amphotericin B
Hate	**H**ypertension	
Dogs	**D**iabetes Mellitus, **D**rugs*	
Mating	**M**isc. → Sarcoid, Calcium, HIV	

Hematuria

I	Idiopathic	**INEPT GUN** Differential diagnosis for **hematuria**
N	Neoplasm*	
E	Exercise	
P	Polycystic kidney disease	
T	Trauma	*malignancy of bladder, kidney, prostate.
G	Glomerular dz (nephritic, nephrotic)	
U	Urinary tract infection	
N	Nephrolithiasis	

Nephrotic Syndrome

T	**T**umor (NHL → minimal change dz)	**THIS LAD HAS** Nephrotic Syndrome Differential diagnosis for **Secondary Causes** of Nephrotic Syndrome. (> 3.5 g/day)
H	**H**eroin* (FSGS)	
I	**I**nfection (see below)	
S	**S**ystemic (see below)	
L	**L**upus (Membranous)	
A	**A**myloid (Membranous)	
D	**D**iabetes (Kimmelsteil-Wilson lesions)	*other drugs gold, penicillamine
H	**H**epatitis B (Membranous, PAN) **H**epatitis C (MPGN)	
A	**A**IDS (HIV-assoc nephropathy, FSGS)	
S	**S**yphilis (Membranous)	

*Pearls: Clinical Clues on Renal U/S:
Small kidneys = Chronic renal disease.
(If unilateral, think of renal artery stenosis)
Large kidneys = Diabetes, HIV nephropathy, Amyloid, Infiltrative dz, PCKD

Urinary Tract Infection

S	*Serratia marcescens*
E	*Escherichia coli*
E	*Enterobacter cloacae*
K	*Klebsiella pneumoniae*
S	*Staphylococcus saprophyticus*
P	*Proteus mirabilis*
P	*Pseudomonas aeruginosa*

SEEKS PP
Most common **bacterial organisms** causing **UTIs.**

Chapter 5: Neurology

A. Neuro Clinical Pearls

Encephalopathy, Hepatic: Precipitating Factors

- Alkalosis
- Budd Chiari
- Constipation
- Drugs (sedatives)
- GI bleed
- Hypokalemia ↓ K^+
- ↑Increased protein diet
- Noncompliance with lactulose
- Renal failure
- SBP – Infection

Korsakoff Dementia

- Should be at the top of the differential diagnosis if patient presents with **confabulation** and **memory loss**.

Neuro Tracts: Crossing and Columns

Neuro Tract	Loss	Deficits occur
- Spinothalamic	✗ Pain and Temp	- Contralateral - Below cord lesion
- Proprioception	✗ Stereognosis	- Ipsilateral - Below cord lesion
- Spinocerebellar	✗ Balance	- Ipsilateral
- Corticospinal	✗ Motor	- Ipsilateral - Below cord lesion

Seizure: Status Epilepticus

Definition	- 1 or more sequential seizures **without** full recovery of consciousness---OR-- **5 MIN** or more of a continuous seizure.

Etiology	- Noncompliance with Anti-epileptic drugs [AEDS]
	    - EtOH withdrawal or OD - Bleed
	    - Benzodiazepine withdrawal - Tumor
	    - Cerebrovascular accident - Meningitis
	    - Electrolyte disturbance

Work-Up	☑ Alcohol level ☑ CBC ☑ Antiepileptic drug levels ☑ CT/MRI ☑ Electroencephalograph(EEG) ☑ Glucose ☑ Lumbar puncture ☑ Prolactin ☑ CMP ☑ TSH ☑ UDP
Complicat.	- Lactic Acidosis - Hyperkalemia - Fever - Hypoglycemia - Pneumonia (aspiration) - Hyponatremia - Rhabdomyolysis

Spinal Cord Injuries – Clinical Management

- PE shows evidence of **neurological deficits**. Next steps:
  - ☑ Need to **start Dexamethasone**
  - ☑ Consult Neurosurgery
  - ☑ Consult possibly with Radiation Oncology

  *Pearl: The most important prognostic sign is the degree of neurologic impairment at presentation.

Stroke, Causes of Stroke other than Atherosclerosis or Emboli

Inflammatory conditions	- Primary granulomatosis angitis - SLE - Temporal arteritis - Takayasu arteritis
Primary hematologic abnormalities	- Anticardiolipin syndrome - Factor V Leiden - G6PD deficiency - Hyperhomocysteinemia - Hereditary spherocytosis - HIT - Sickle cell anemia - TTP - Thalassemias - PT 20210 mut. - PNH
Others	- Migraines - Vasospasm in the setting of subarachnoid hemorrhage - Venous sinus thrombosis

Stroke, Clinical Lacunar Syndrome and Infarct Location

Clinical Syndrome	Location of Lacunar Stroke
Pure **motor** hemiparesis (face, arm, leg)	- Contralateral posterior limb of **internal capsule**
Pure unilateral **sensory loss** (face, arm, leg)	- Contralateral **thalamus**
Hemiparesis + Ataxia	- Contralateral **thalamocapsular** region or upper 1/3 on contralateral medial **pons**.
Clumsy hand, dysarthria	- Contralateral lower 1/3 of medial **pons**.

Wernicke's triad

Take Home Point: Triad of symptoms often presented in alcoholic pts.
1. Ophthalmoplegia
2. Ataxia
3. Confusion

*Pearl: Remember to give **Thiamine** (Vit B1) ***BEFORE*** glucose.

B. Neuro Differential Diagnosis

Cerebellar Ataxia: DDx

Infectious	<ul><li>EBV</li><li>Enterovirus</li><li>Creutzfeldt-Jakob</li><li>Lyme</li><li>Hep A</li><li>HSV 1</li><li>HSV 6</li><li>Parvo B19</li></ul>	<ul><li>Measles</li><li>Mumps</li><li>Mycoplasma</li><li>Rubella</li><li>Typhoid</li><li>Vaccination</li><li>Varicella</li></ul>
Immune/ Inflammatory	<ul><li>ADEM (Acute demyelinating encephalomyelitis)</li><li>Multiple Sclerosis</li></ul>	
Neoplastic	<ul><li>Primary tumor vs. mets</li></ul>	
Toxic/Metab.	<ul><li>Thiamine deficiency</li><li>Vitamin E deficiency</li><li>Ethanol</li><li>Anticonvulsants</li></ul>	<ul><li>Narcs, Benzos</li><li>Chemo</li><li>Mercury</li><li>Lithium</li></ul>
Vascular	<ul><li>Hemorrhage</li><li>Infarct</li></ul>	<ul><li>Venous</li><li>Thrombosis</li></ul>

Facial Nerve Palsy: DDx

- Bell's palsy
- CNS lymphoma
- Guillain-Barré syndrome
- Lyme disease
- Malignant otitis media
- Melkersson-Rosenthal
- Meningeal carcinomatosis
- Ramsay Hunt syndrome
- Sarcoid
- Syphilis
- Talosa-Hunt syndrome
- Trauma
- Tuberculosis meningitis
- Tumor

Peripheral Neuropathy, Painful: DDx

- Amyloid
- Alcoholism
- Diabetes mellitus
- Drugs
- Fabry's
- Heavy metal
- HIV
- Myeloma
- Paraneoplastic
- Sarcoid
- Tabes dorsalis
- Uremia
- Vasculitis

Stroke in the Young: DDx

- Aneurysm/AVM
- Cocaine
- Dissection
- HIV
- HSV
- Hypercoag state (ex. APL-LA)
- Migraine
- Multiple sclerosis
- Sarcoidosis
- Seizure with subsequent Todd's paralysis
- Syphilis
- Trauma
- Vasculitis
- Systemic lupus erythematosus
- Polyarteritis nodosa

Transverse Myelitis: DDx

- Behçet's
- Neoplastic, mass lesions
- Paraneoplastic
- Sarcoid
- Sjögren's syndrome
- Vascular
- Osteomyelitis
- Rheumatoid arthritis
- Tuberculosis

Infectious Causes
- Cat scratch
- CMV
- Enterovirus
- HSV 1, HSV 2, HSV-6
- HTLV – 1
- Lyme disease
- Mycoplasma
- VZV

C. Neuro Mnemonics

Altered Mental Status

D	**D**rug or drug withdrawal	**DELERIUMS**
E	**E**lectrolytes (↓ Na⁺ and ↑ / ↓ Glu)	DDx for acute change in mental status.
L	**L**iver (Hepatic encephalopathy)	
	Lung (Pneumonia)	
E	**E**ndocrine*	* Thyroid storm, Addison's disease, Hypercalcemia, Myxedema coma
R	**R**enal/Uremia	
I	**I**nfection/Sepsis	
U	**U**nstable (MI, ↓BP, Pul edema, CVA)	
M	**M**etabo. (hypoxia, ↓osmolar, acidosis)	
S	**S**eizure/Stroke/Subdural/Subarachnoid	

Dementia

D	**D**egenerative (Parkinson, Huntington's)	**DEMENTIASS**
E	**E**ndocrine (Thyroid, Pit., Parathyroid)	Major etiologies of **Dementia**.
M	**M**etabo. (ETOH, Glu, 'Lytes, Hepatorenal)	
E	**E**xogen. (CO pois., drugs, heavy metal.)	*Infx causes: encephalitis, meningitis, cerebral abscess, syphilis, prions, HIV, Lyme disease.
N	**N**eoplastic	
T	**T**raumatic	
I	**I**nfectious	
A	**A**ffective disorders (pseudodementia)	
S	**S**troke (multi-infarct, ischemia, vasc.)	
S	**S**tructural (NPH)	

Horner's Syndrome

A	**A**nhidrosis	**AMP**
M	**M**iosis	Signs/Symptoms of **Horner's** syndrome.
P	**P**tosis	

Normal pressure hydrocephalus

W	**W**acky (cognitive impairment)	**3Ws**
W	**W**et (incontinence)	Signs of **Normal pressure hydrocephalus**.
W	**W**obbly (gait abnormalities)	

Parkinson's Disease

S	**S**huffling gait	**SMART**
M	**M**ask-like facies	Signs of **Parkinson's disease**.
A	**A**kinesia	
R	**R**igidity ("cogwheel")	
T	**T**remor (resting)	

Peripheral Neuropathy

D	Diabetes mellitus	
D	Drugs (Thalidomide, Arava, Pyridoxine)	
A	Alcohol	
A	Acromegaly	
N	Nutritional B12	
G	Guillain Barré syndrome	
G	Gammopathy	
T	Toxin	
H	Hereditary	
E	Entrapment	
E	Endocrine	
R	Refsum's disease	
R	Rheumatoid	
A	Amyloid	
P	Porphyrias	
P	Polyarteritis	
P	Polycythemia vera	
P	Polymyositis	
P	Paraproteinemia	
I	Infection	
I	Inflammatory	
S	Systemic disease	
T	Tumor	

Dang Therapist
Causes of **peripheral neuropathy**

*Other causes:
Cryoglobulinemia,
Churg-Strauss synd,
Sjögren's synd,
Paraneoplastic synd,
Charcot-Marie-Tooth dz,
Syphilis, Leprosy.

Subarachnoid Hemorrhage

B	Berry aneurysm
A	Arteriovenous malformation
T	Trauma
C	Cortical thrombosis
A	Angioma
V	Vasculitis
E	Embryologic abn – (coarc Aorta, Marfan, APKD, Ehlers-Danlos, fibromuscular dysplasia)
S	Smoking (risk factor)

BAT CAVES
Causes of **SAH**.

Pearls: SAH H&P Pertinent (+'s)
→ Age ≥40
→ Neck pain /stiffness/↓flexion,
→ Loss of consciousness
→ Onset of HA during exertion,
→ Thunderclap headache
→ OCP use, hormone replac.
→ Vigorous exercise

Stroke, Risk Factors

H	Hypertension	**HEADACHES**
E	Elderly	Risk factors for **stroke**.
A	Atrial fibrillation*	
D	Diabetes mellitus	*For **Afib** see
A	Atherosclerosis	Atrial Fibrillation
C	Cardiac defect	Stroke Risk:
H	Hyperlipidemia	$CHA_2DS_2VAS_c$
E	Excess weight	Score.
S	Smoking	

Stroke, Middle Cerebral Artery

C	Contralateral paresis and sensory loss in the face and arm	**CHANGe** Middle cerebral artery stroke signs and symptoms.
H	Homonymous hemianopia	
A	Aphasia (dominant)	
N	Neglect	
G	Gaze preference toward lesion side	

Stroke, Posterior Circulation

D	Diplopia	**4 Deadly Ds**
D	Dizziness	Posterior
D	Dysphagia	circulation stroke
D	Dysarthria	signs/symptoms.

Stroke, Causes in Young Patients

C	Cocaine	**7 Cs**
C	Cancer	Causes of **stroke**
C	Cardiogenic emboli	**in young patients.**
C	Coagulation (excessive)	
C	CNS infection (septic emboli)	
C	Congenital vascular lesion	
C	Consanguinity (genetic disease)	

Chapter 6: Infectious Diseases

A. ID Clinical Pearls

Antibiotics: Classic Side Effects

Antibiotic	Side Effects
Vancomycin	- Red Man syndrome
Metronidazole	- Metallic taste - Disulfiram-like reaction - *Pearl*: Patients should avoid alcohol
Nafcillin	- Acute interstitial nephritis
Clindamycin	- Pseudomembranous Colitis
Ciprofloxacin	- Cartilage damage
Trimethoprim-sulfamethoxazole	- Rash Hyperkalemia - Stevens-Johnson syndrome - Pseudo-elevation in Creatinine.
Macrolide	- GI upset QT prolongation
Tetracycline, Doxycycline	- Photosensitivity Teeth discoloration
Aminoglycoside, Tobramycin	- Nephrotoxicity Ototoxicity
Penicillin Cephalosporin 1st gen	- Hypersensitivity Reaction - *Pearl*: If ↑ dose IV PCNs risk for volume overload, AIN, Hyperkalemia
Minocycline	- Dizziness (especially in females)

Clostridium Difficile: New Strain

Key Features	- NAP1 strain – hypertoxigenic strain - Binary toxin with ↑ **increased virulence** - Assoc/w fluoroquinolone use +/- Cephalosporins
Top 3 abx a/w C. Difficile	- Fluoroquinolones -Cephalosporins - Clindamycin *Pearl*: It can be <u>any</u> antibiotic.
Pearl	Other ABXs can cause antibiotic associated diarrhea unrelated to Clostridum Difficile.

Enterococcus: Sources of Infection and Treatment

Sources:
- Biliary
- GI
- Endocarditis
- Urine

Tx: Depends on sensitivity
- ☑ Ampicillin OR Vancomycin

Ehrlichiosis, Clinical Picture

Clinical:
- Abd. pain
- Confusion
- Myalgia
- Fatigue

Labs
- ☑ Lymphocytic pleocytosis
- ☑ Leukopenia
- ☑ ↑LFTs

Fever of Unknown Origin: DDx, Pearls, Work-Up	
Etiology	- Neuroleptic malignant syndrome - Serotonin syndrome ☑ Pearl: <u>always</u> check medication list
Cardiovascular Disease	- Adult Still's disease - Rheumatic fever/Rheumatic heart disease - Systemic lupus erythematosus - Vasculitis (ex. Temporal arteritis)
Infection	- Bacterial endocarditis - Osteomyelitis - Occult abscess - Tuberculosis
Tumors	- Hepatoma - Myxoma - Lymphoma - Renal
Medications	- Dilantin - Sulfa - Lasix
Other Causes	- Addison's disease - Hyperthyroidism - Cocaine - Septic pelvic thrombophlebitis - Factitious
Rare Causes	- Macrophage activation syndrome

Fever of Unknown Origin

Summary – A stepwise approach to the work-up for a fever of unknown origin, depending on the clinical scenario.

FUO Def. = >101 F (38.3 C), > 1 episode, >/= 3wks, No DX s/p 1wk workup.

Diagnostic Level	Work-Up
Level 1	☑ Blood cultures x 3 ☑ CBC with differential ☑ CMP ☑ Chest X-Ray ☑ ESR and C-reactive protein (CRP) ☑ Liver function tests ☑ Lower extremity Doppler's ☑ Purified protein derivative (PPD) ☑ Rapid plasma reagin (RPR) ☑ Urinalysis
Level 2	☑ ANA ☑ Fungal infection [histoplasmosis antigen test] ☑ Viral [Echovirus, EBV, CMV, HIV] ☑ Rheumatoid factor
Level 3	☑ Bone marrow biopsy ☑ CT Scan – Thorax, Chest, Abdomen [consider] ☑ Gallium scan ☑ Lumbar puncture ☑ Trial tuberculosis medications ☑ Triple phase bone scan

Fever, Causes for Fever in Intensive Care Unit (ICU)

- Acalculous cholecystitis
- Addison's disease
- Alcohol withdrawal
- Blood transfusions
- Check all IV sites
- Cholelithiasis
- CNS - meningitis
- Dissected aortic aneurysm
- Dressler's syndrome
- Drug fever
- Deep venous thrombosis
- Gout
- Hepatitis
- Infections
- Ischemic colitis
- Myocardial infarction
- Pancreatitis
- Pulmonary embolism
- Pericarditis
- Vasculitis

Fever, *Last Ditch* Workups to Verify a Cause

Take Home Message: If standard work-up for source of fever / fever-of-unknown origin have been exhausted and were unremarkable, here are the work-ups that are your last resort to try and determine source of fever.

- Bone marrow biopsy
- CT Body
- Dukes criteria for endocarditis
- Lower extremity doppler
- Liver biopsy
- Temporal artery biopsy

Clinical Pearl: STOP any offending drugs.

Lemierre's Syndrome

- Postanginal sepsis caused by ***Fusobacterium* necrophorum:** is a septicemia resulting from an antecedent **pharyngitis** that causes an **internal jugular vein thrombophlebitis**.
- Caution for subsequent **septic emboli**: treat infection promptly!

Meningitis and Encephalitis: Etiologies

Bacterial Meningitis		Viral Meningitis	
Strep pneumo	N. meningitidis	HSV	Arboviral
H. Influenzae	Group B strep	West Nile virus	Equine
Listeria monocytogenes		Enterovirus	Mumps

Meningitis: Cerebrospinal Fluid Analysis Workup

- ☑ Cell count and differential
- ☑ Gram stain
- ☑ Glucose and Protein
- ☑ Culture and sensitivity
- ☑ Viral cultures
- ☑ Acid-fast bacillus
- ☑ Fungal stains and culture
- ☑ VDRL
- ☑ Cryptococcus antigen

Nosocomial Infection, *5 Bugs to Keep in Mind*

1. MRSA
2. E. Coli
3. Staph Epidermidis
4. Enterococcus
5. Pseudomonas

Sepsis: Hx, Etiology, Gram Negatives	
Etiology Pearls	<ul><li>Identify source: Lung, Urine, GI, Catheter sites, Skin, and Soft tissue</li><li>Immunocompromised – Fungal, Viral, Parasitic</li><li>*Caution* with pelvic exam</li><li>Source Unknown ~ 30% of cases.</li></ul>
Gram Negative Causes	<ul><li>*E. Coli*</li><li>*Pseudomonas spp.*</li><li>*Enterobacter spp.*</li><li>*Klebsiella spp.*</li><li>*Proteus spp.*</li><li>*Providencia spp.*</li></ul>
HPI Pertinent Positives	**General Symptoms:** Wt loss, diarrhea, rash, abd pain **Musculoskeletal:** Stiff neck, back pain **HEENT:** Cough, sputum, sore throat, poor dentition **Chronic Disease:** HIV, malignancy, tuberculosis, hepatitis **Social:** Travel, pets, insect bites, sick contacts, **Genitourinary:** Prostate prob., vaginal discharge, dysuria
Other DDX	<ul><li>Addison's disease</li><li>Alcohol withdrawal</li><li>Bacterial endocarditis</li><li>DVT</li><li>Drug fever</li><li>Giant cell arteritis</li><li>Takayasu's arteritis</li><li>Gout</li><li>Hypothyroidism</li><li>IBD</li><li>Leukemia</li><li>Lymphoma</li><li>MI</li><li>Pancreatitis</li><li>Pulmonary embolism</li><li>Rheumatic fever</li><li>SLE</li></ul>

West Nile Virus: Signs and Symptoms

- Acute flaccid paralysis (poor prognosis)
- Encephalitis
- Meningitis
- Myoclonus
- Neuro manifestations
- Parkinsonism
- Polio-like anterior horn cell dysfunction
- Upper extremity tremor

B. ID Differential Diagnosis

Bacteremia, Gram Negative: DDx

- Bacterial endocarditis
- Gonococcal infection
- Colonic flora: Localized intestinal infection (ex. appendicitis and diverticulitis)
- Co-infx w/ *Strongyloides sterc.*
- Melioidosis
- Typhoid illness
- Urinary tract infection
- Vibrio species

HIV / Lymphadenopathy / Pulmonary Infiltrates / Cytopenias: DDx

- Bartonella
- Castleman's disease (HHV-6)
- Fungal
- Nocardia
- Rhodococcus
- Syphilis
- Tuberculosis / Mycobacterium avium-intracellulare

Peripheral Lymphadenopathy: DDx

Infections	- **Bacterial:** abscess, cellulitis, sinusitis, otits externa, tonsillitis, pharyngitis, cat-scratch disease - **Viral:** EBV, CMV, HIV - **Fungal:** Sporotrichosis, Cryptococcus - **Mycobacterial:** TB - **Parasite:** Toxoplasmosis
Immunologic	- Sarcoidosis - SLE
Neoplasms	- Non Hodgkins and Hodgkin lymphoma - Breast cancer (Axillary) - Head & neck cancer (Cervical)
Misc.	- Castleman disease - Kikuchi-fujimoto ds (histiocytic necrotizing lymphad.)

Meningitis and Encephalitis, Historical Clues to Etiology DDx

- Alcoholism
- Antibiotic use
- Diabetes mellitus
- HIV
- Intravenous drug abuse
- Mosquito bite
- Rash
- Pets
- STDs
- Sick kids:
  - Contacts
  - Recent URI
  - Ear aches
  - Sinusitis
- Swimming in Lake/Pond
- Tick exposure
- Travel

C. ID Mnemonics

Endocarditis

H	*Haemophilus*	**HACEK BB**
A	*Actinobacillus*	Endocarditis
C	*Cardiobacterium*	culture negative
E	*Eikenella*	organisms
K	*Kingella*	
B	*Bartonella quintana*	
B	Coxiella *Burnetii*	

Post-Operative Fever

Wind	Pneumonia, Atelectasis	**5 Ws of Post-Operative Fever**
Water	UTI	Major causes of **fever in post-op patients.**
Wound	Wound infection	
Wonder drugs	Anesthesia, Drug fever	
Walking	Or else risk DVT, Pulmonary embolism	

D. ID Clinical Cases #1

History
- 40-year old male presents to the ED with c/o left knee pain and swelling for one week. PE reveals moderately large effusion, some warmth, not much erythema. Pt denies trauma. Upon review of systems pt denies presence of penile drip, no history of STIs, no arthritis, no murmur, and no history of Gout. Pt does have low-grade fever.

Objective Findings:
- **Skin:** Large targetoid appearing rash, lesion on back.
- **CV:** No murmurs
- **Lungs:** Clear to auscultation bilaterally

Laboratory Findings:
Urinalysis	Within Normal Limits

Arthritis, Monoarticular: Differential Diagnosis:
- Degenerative joint disease
- Endocarditis
- Gonococcal
- Gout
- Reactive / Reiter's
- Sarcoidosis
- Septic joint
- Trauma
- Lyme disease
- Pseudogout

Final Diagnosis:
- Lyme Disease

Clinical Summary – Lyme Disease
- Musculoskeletal manifestations of Lyme disease are very common. During early infection, migratory arthralgias and pain in bursae, tendon, muscle, or bone occur in the majority of patients. Weeks to months later, frank arthritis, most commonly mono- or oligoarticular and involving large joints (most commonly knees, but also shoulders, ankles, elbows, and other sites), may develop. **Lyme arthritis** is one manifestation of **persistent** or **late Lyme disease**.
- **Stage 1** (early): Flu-like likeness, **erythema migrans** 2° efx spirochete
- **Stage 2** (weeks-months): fatigue, malaise, HA, **mult annular lesions, CN 7 neuropathies, heart block** 2° efx spirochete/immune resp.
- **Stage 3** (late persistant): recurrent **mono-/oligoarthritis large joints, acrodermatitis chronica atrophicans** 2° efx chronic infx/autoimmune.
- Infection with spirochete **Borrelia burgdorferi** – dx Elisa +/- West. Blot
- High risk regions: **NY, NJ, CT, RI, WI, PA, MA, ME, NH, MI, MD, DE**
- Source: http://www.hopkins-arthritis.org/arthritis-info/lyme-disease/clinical-presentation.html#target

D. ID Clinical Cases #2

History:

History of Present Illness:
- 19-year old male college student presents with sore throat, fever, chills, nausea, vomiting, and right sided pleuritic chest pain. Over the previous 2 weeks he had noted a seven pound weight loss and a swelling over the right side of his neck.

Past Medical History:
- History of mononucleosis 2 years ago. He takes no medications.

Social History:
- He denies intravenous drug use. Patient has had many heterosexual experiences over the past year, and admitted to one homosexual contact one month prior to admission.

Vitals and Physical Exam:

T: 102.5° F HR: 110 beats/min RR: 26 breaths/min BP: 108/80

- ☑ **HEENT**: Erythematous pharynx w/no exudates. Right anterior neck is warm and tender.
- ☑ **CV**: No murmur
- ☑ **PULM**: Moderate dyspnea. Bibasilar crackles in all lung fields.
- ☑ **ABD**: Palpable spleen
- ☑ **Skin**: No rash

Laboratory Finding	
WBC	WBC = 16,000 • 82% PMN, 16% Lymphs, 2% Monos
Hb	13
Platelets	Normal
BMP	Normal
LDH	341
ABG [on room air]	7.46 / 36 / 59
Sputum	Mixed flora
Chest X-Ray	Left lower lobe pneumonia / effusion

Next Steps in Management:
- Patient was started on broad spectrum intravenous antibiotics.
- The next day, 2 of 2 anaerobic blood cultures were positive for a gram (-) negative rod.

Clinical Question: What syndrome is suggested by the patient's presentation and culture results?
- **Lemierre's syndrome**

Clinical Pearls: Lemierre's syndrome

- **Lemierre's syndrome** is characterized by **disseminated abscesses and thrombophlebitis of the internal jugular vein** after infection of the **oropharynx**. The predominant pathogen is a gram-negative anaerobic bacillus, ***Fusobacterium necrophorum***.
- The usual cause is *Fusobacterium necrophorum*, but **Streptococcus, Bacteroides, Peptostreptococcus,** and **Eikenella** are also causes.
- Necrobacillosis and post-anginal sepsis are old terms.

D. ID Clinical Cases #3

History:

- 24-year old female presents to your office with 5 days of fevers, lymphadenopathy, and sore throat. Patient also has fatigue and vague abdominal pain.

Laboratory Findings:

Liver Function Tests	AST = 80 units/L ALT = 90 units/L ALK PHOS = 170 units/L
Total Bilirubin	1.9 mg/dL
WBC	11
Hb	11
Platelets	150
Peripheral Smear	Atypical lymphocytes

Differential Diagnosis for Atypical Lymphocytes:

- Cytomegalovirus
- HIV
- Toxoplasmosis
- Cutaneous T Cell Lymphoma
- Mononucleosis (Epstein-Barr virus)

Final Diagnosis:

- **Mononucleosis** (Epstein-Barr virus).

Clinical Pearls – Mononucleosis

- Primary **HIV** infection can resemble mononucleosis, **acute EBV infection**; patients with **risk factors for HIV infection** should be tested using **quantitative HIV RNA viral count** and **p24 antigen assay**
- ELISA/Western blot tests are usually negative during **acute HIV infection**.

Chapter 7: Endocrine & Metabolism

A. ENDO/METABO Clinical Pearls

Adrenal Incidentaloma: Pearls and Workup

Etiology:	- Adrenal masses are common ↑ increase in freq with age. - May be **benign** (non-functional), **malignant** (primary or mets) or **functional** (Pheo, Conn's, Cushing's)
Size and Character	- Masses < 3cm are rarely functional tumors. Can be followed with serial imaging. - Masses **< 4cm** = 2% chance of cortical carcinoma - Masses **4-6cm** = 6% chance of cortical carcinoma - Masses **> 6cm** = 25% chance of cortical carcinoma - If adrenal mass is large, biopsy it. Definitely do so if **> 6cm** after ruling out Pheo. - CT characteristics - **<4cm and <10 HU** (Hounsfield Units), homogenous with smooth border = **very likely benign**.
Clinical Challenge	Work-up an adrenal incidentaloma - if it is functional?
	Cushings Syndrome	☑ Dexamethasone suppression test
	Pheochromocytoma	☑ Plasma free metanephrines
	Conn's Syndrome	☑ Only check IF pt is hypertensive ☑ Check aldo/renin ratio

Cushing's Syndrome

Clinical Picture:
- Acne
- Round facies
- Hypertension
- Hyperglycemia
- Hypokalemia
- Metabolic alkalosis
- Purple stria
- Truncal obesity

Workup
- ☑ 24-hour urine free cortisol
- ☑ Dexamethasone suppression test
- ☑ Salivary cortisol levels
   [*Pearl: <u>Time</u> of day is important]

Diabetes Mellitus Type 2 Medications: Black Box Warning

Metformin
- <u>DO NOT</u> give to patients with **renal insufficiency**.
  - ✗ Serum Cr > 1.4 in females
  - ✗ Serum Cr >1.5 in males
- May increase risk of **lactic acidosis**.
- <u>STOP</u> Metformin the **day of contrast/procedure/surgery**, and resume in **48 hours**.

Thiazolidinedione
- <u>DO NOT</u> give to **HEART FAILURE** patients.

Hyperglycemia ↑ Blood Glucose: Secondary Causes	
Endocrine	- Acromegaly - Cushing's syndrome - Disorders of exocrine pancreas - Hyperthyroidism - Islet cell neoplasms (Glucagonoma, Somatostatinoma) - Pheochromocytoma
Medications	- Glucocorticoids
Pregnancy	- Gestational diabetes

Hyperkalemia ↑ K^+: Major Causes

General causes of ↑ K^+
- Acidosis
- Adrenal insufficiency
- Digoxin toxicity
- Renal failure
- Renal tubular acidosis - Type IV
- Rhabdomyolysis
- Tumor lysis syndrome

Medications that cause ↑ K^+
- ACE Inhibitors
- Bactrim
- Beta-blockers
- Heparin
- K^+ sparing diuretics (Aldactone)
- NSAIDs

Hyperthyroidism, Causes & Pearls

Causes of Hyperthyroidism
- Factitious
- Graves' Disease
- Struma ovarii
- Thyroiditis
- Amiodarone / Wolff-Chaikoff Effect
- Iodine-induced hyperthyroidism / Jod-Basedow phenomenon
- Multi or single nodular goiter

Pearls
- Iodine induced hyperthyroidism called the **Jod-Basedow phenomenon.**
- Amiodarone Induced Thyroid disease, called the **Wolff-Chaikoff Phenomenon**. Patient can have hypothyroidism or hyperthyroidism in Wolff-Chaikoff.

Hypocalcemia, Causes of ↓ Ca^{2+}

- Alcoholism
- Decreased ↓ Magnesium
- Deficiency in ↓ Vitamin D
- Decreased ↓ Phosphate
[*Pearl: decreased phosphate is mainly caused by renal failure]
- Extensive transfusions /sepsis
- Malabsorption (Vit A, D, E, K)
- Pancreatitis
- Primary hypoparathyroidism

Lactic Acidosis	
A type	- **Hypoperfusion** due to ↓low cardiac output states (i.e., shock, severe anemia, cardiac arrest) - Most common cause.
B type	- Metformin - Malignancy (Lymphoma, Leukemia) - HIV meds (NRTI's)
D type	- Short bowel syndrome/Malabsorption (especially after high carbohydrate meals)
- **Other Causes of lactic acidosis**: CO poisoning, severe asthma, COPD, asphyxia, regional hypoperfusion (limb or mesenteric), DKA	

MEN Syndromes	
MEN I (Wermer's Syndrome):	MEN 2A (Sipple's Syndrome):
1. Pancreatic / Gastrinoma 2. Pituitary adenoma 3. Parathyroid disorders	1. Medullary cancer of the thyroid 2. Pheochromocytoma 3. Parathyroid disorders
	MEN 2B:
Mnemonic – **MEN 1 affects "P" organs**	1. Medullary cancer of the thyroid 2. Pheochromocytoma 3. Mucosal neuromas 4. Marfanoid habitus

Thyroid Uptake and Scan Results	
NO Uptake on Scan	Uptake on Scan
- Thyroiditis - Factitious - Iodine induced	o Graves o Multinodular goiter o Hyperfunctioning nodule

Thyroid Medication, Pearls	
HYPOthyroidism	**HYPER**thyroidism
- Thyroid hormone meds need to be ↑ **increased** during pregnancy ~ 30% - 40%.	o PTU/Methimazole ok to stop 3-7 days prior to Iodine-131 Ablation o **Avoid** pregnancy for 6-12m

Whipple's Triad

Take Home Points: If patient has the 3 diagnostic criteria of Whipple's Triad→ patient's symptoms result from **hypoglycemia**.

1. **CNS Changes**
2. Accu-Cheks show ↓ **Blood Glucose** (BG) =/< 50mg/dL
3. Symptoms **RESOLVE** with **glucose administration**

B. ENDO/METABO Differential Diagnosis

Hypoglycemia: DDx

- Adrenal insufficiency
- Beta blockers
- Ethanol metabolism blunts gluconeogenesis
- Factitious
  o Surreptitious insulin
  o Oral hypoglycemic use
- Hepatic failure (advanced)
- Sepsis
- Sarcomas – large retroperitoneal sarcomas
- Sulfonylureas
- Tumors
- Islet beta cell tumors (pancreatic) insulinomas
- Non-islet cell tumor: Large mesenchymal tumors
- Uremia / renal failure

Hypokalemia, Metabolic Alkalosis, and HTN: DDx

- Aldosteronism/Primary Hyperaldosterone State
- Conn's syndrome
- Bilateral adrenal hyperplasia
- Cushing's
- Deoxycorticosterone producing tumor (DOC tumor - adrenal)
- Diuretics (BP may be normal)
- Licorice and chewing tobacco due to glycyrrhetinic acid induced-pseudoaldosteronism
- Liddle syndrome
- Malignant hypertension
- Renal artery stenosis
- Scleroderma crisis

Hyponatremic, Euvolemic: DDx

- Adrenal insufficiency
- Beer potomania
- Hypothyroid
- Psychogenic polydipsia
- SIADH hormone hypersecretion
- Tea and toast syndrome

*Pearl: Cerebral salt wasting is indistinguishable from SIADH on urine lytes. Distinguishing feature is that cerebral salt wasting patients are **volume depleted**.
*Pearl: SIADH → low or normal **uric acid level**.

Hypophosphatemia: DDx

- Alcoholism
- Burns
- Decreased Vitamin D
- Diuretic
- Diabetic ketoacidosis
- Hyperaldosteronism
- Malabsorption
- Malnutrition
- Phosphate binders
- Refeeding syndrome
- Renal tubular acidosis [Type 2]

C. ENDO/METABO Mnemonics

Acidosis, High Anion Gap

A	**A**lcoholic ketoacidosis	A MUD PILES
M	**M**ethanol (ex. fuel, antifreeze: Caution can cause **blindness!**)	Causes of ↑ High Anion Gap Acidosis
U	**U**remia	
D	**D**iabetic ketoacidosis (*Pearl: ketones)	
P	**P**ropylene glycol	
I	**I**soniazid/Iron	
L	**L**actic acidosis	
E	**E**thylene glycol* (radiator, antifreeze, windshield deicer, windshield wiper fluid.)	* Ethylene glycol pearl: Oxalate
S	**S**alicylates (s/s GI upset/bleed, tinnitus)	Crystals in urine.

Pearls:
- Top 2 Causes are **Diabetic ketoacidosis** and **Lactic acidosis**
- *Anion Gap* is endogenous production of organic acids.
- *NON-anion gap* is due to **loss of bicarbonate**: Usually due to **bowel (diarrhea)** or **kidney (RTA) problems**.
- Every Rise in ↑ Anion Gap by **1** should decrease ↓ HCO3 by **1**.

Hypercalcemia, Causes

M	**M**alignancy (Thyroid, Breast, Lung)	MISHAP
I	**I**ntoxication (Vitamin D, Vitamin A)	Main Causes of
S	**S**arcoid (RTA – distal) (granulomas)	Hypercalcemia
H	**H**yperparathyroid	
A	**A**lkali (Milk alkali syndrome)	
P	**P**aget's disease	

Pearls:
- 90% of hypercalcemia is due to either PHPTH or malignancy.
- 80% of causes of PHPTH is due to **single functioning adenoma** and hyperplasia causes the other 20%.
  - **Other Causes of Hypercalcemia:** Multiple Myeloma, Lithium, Pheo, TB, Crohn's, Familial Hypocalciuric Hypercalcemia, diuretics, endocrine (Addison's disease, thyrotoxicosis).

Hypercalcemia, Complications

Stones	▪Renal calculi – nephrolith., nephrocalc.	Stones, Bones,
Bones	▪Osteitis fibrosa cystica, bone cysts, "Brown tumor", osteoporosis	Abdominal moans, and
ABD moans	▪Constipation, acute pancreatitis, PUD	Psych
PSYC overtones	▪Altered level of consciousness, psychosis, depression, personality changes.	Overtones Complications of Hypercalcemia

Hypoglycemia

Nutr	Prolonged starvation (pre-anesthesia, protein calorie malnutrition, ↓low-calorie ketogenic diet), renal dz.	Not only **NUTRITION EXPLAIN**s it. Causes of **Hypoglycemia**.
EX	**EX**ogenous drugs*	
P	**P**ituitary insuff. (↓GH or cortisol) **P**sychogenic	***Drugs** - Insulin, oral hypogly., EtOH, cocaine, salicylates, beta-blockers, pentamidine
L	**L**iver disease**	
A	**A**drenal failure (↓ cortisol) **A**utoimmune (Grave's dz)	***Liver dz** - hepatic failure, cirrhosis, galactose intol., fructose intol., glycogen stor. dz.
I	**I**nsulinomas **I**mmune hypoglycemia	
N	**N**on-pancreatic neoplasms (retroperitoneal sarcoma)	

*Other Causes of Hypoglycemia:** Addisons, glucagon deficiency, Cardiac dysrhythmia, CNS disorders, TIA

Pheochromocytoma

5 Ps and Rule of 10s

P	Pressure	**10%**	Extra-adrenal
P	Pain (headache)	**10%**	Malignant
P	Perspiration	**10%**	Bilateral
P	Palpitations	**10%**	Familial
P	Pallor	**10%**	NOT assoc/w HTN

Symptoms & Factors of **Pheochrom.**

D. ENDO/METABO Case #1

History:
- 33-year old male with Diabetes Mellitus Type 1 for 29 years, complicated by retinopathy, neuropathy, and nephropathy presented with nausea and vomiting. He stated he had been taking Reglan for presumed gastroparesis with no relief. Patient complained of a 30lb weight loss over 3-4 months. Although blood glucose was under adequate control, he recently required IVF's on several occasions for symptomatic orthostatic hypotension. He was tested for HIV recently, and it was negative.

Objective Findings:

Vital Signs:
- ☑ **Temperature**: 99°F
- ☑ **Pulse**: 100 beats per minute
- ☑ **Respiratory Rate:** 16 breaths per minute
- ☑ **Blood Pressure:** 140/80 (supine), BP = 120/82 (standing)

Physical Exam:
- ☑ **Eyes:** Diabetic retinopathy
- ☑ **Reflexes:** Absent DTR's in knees and ankles
- ☑ **Neuro:** Loss of vibratory and fine touch perception on legs

Laboratory and Diagnostic Findings:	
WBC	5.6 [nl diff]
Hb	12
Platelets	normal
Na+	140
K^+	4.1
Cl^-	107
HCO3	22
Glucose	138
BUN	28
Cr	2.4
Ca^{2+}	14.1
Phos	1.4
Cortisol	24
HbA1C	9.1
Urinalysis	2+ Protein
HIV Test	Negative

Clinical Question

- What diagnosis could explain the patient's symptoms and lab abnormalities?

Hypercalcemia Take Home Points

- 90% of the time it will be these two entities: **1° hyperparathyroidism** or **malignancy**
- **Vitamin D intoxication** occurs when the patient is consuming very large quantities. For instance, more than 150,000 IU/Day.
- Tuberculosis sarcoidosis and other **granulomatous diseases** cause hypercalcemia because macrophages possess an enzyme that transforms 25-hydroxyvitamin D into its highly active metabolite, 1, 25 dihydroxyvitamin D.
- **Milk alkali syndrome** usually occurs from ingestion of large quantities of absorbable antacids that leads to renal failure and alkalosis.
- **Hyperthyroidism** produces mild hypercalcemia by mobilizing calcium from bone.
- **Less common causes include:** Extreme immobilization, Vitamin A toxicity, Paget's disease, Addison's disease.
- **Familial Hypocalciuric Hypercalcemia** is an autosomal dominant disorder that can be confused with PHPT however it is asymptomatic and the PTH is normal. Polyuria and polydipsia due to mild vasopressin-resistant diabetes insipidus. Muscle weakness, lethargy, anorexia, nausea, constipation, and depression can all be present.
- PHPT – 20% can have nephrolithiasis.
- Pancreatitis, PUD, and HTN.
- When HTN is present with <u>PHPT</u> always think of....
- MEN 2a – pheochromocytoma, medullary cancer of the thyroid
    - 80-85 percent of the time PHPT will be due to....
        - A single parathyroid adenoma
        - Almost all other have multiple gland hyperplasia
        - Few have parathyroid cancer
- ***Stones, bones, abdominal moans, and psych overtones*** for complications of hypercalcemia.

Stones	Renal calculi – nephrolithiasis, nephrocalcinosis
Bones	Osteitis fibrosa cystica, bone cysts, "Brown Tumor", osteoporosis
ABD moans	Constipation, acute pancreatitis, PUD
Psychological overtones	Altered level of consciousness, psychosis, depression, personality changes

Chapter 8: Hematology & Oncology

A. HEME/ONC Clinical Pearls

Alpha and Beta Thalassemia

- Target cells on peripheral blood smear = Thalassemias

Alpha thalassemias	# of Gene deletion(s) (4 total)
No defect	☑ 1 gene deletion
Trait (minor)	☑ 2 gene deletion
Hb H disease	☑ 3 gene deletion
Hydrops fetalis (lethal)	☑ 4 gene deletion

Beta thalassemias	# of Gene deletion(s) (2 total)
Trait (minor)	☑ 1 gene deletion
Major (Cooleys)	☑ 2 gene deletion

Anemia, Microcytic: (MCV < 80 fL)

Top 3 Causes:
1. Iron deficiency
2. Thalassemia
3. Anemia of chronic disease

Rare Causes:
- Sideroblastic
- Lead poisoning

Anemia, Macrocytic:

Vit B 12 Deficiency
- Neuro deficits (paresthesias)
- Microscp.: Macro-ovalocytes, Hypersegmented polys

MCV greater than 110 fL – DDx:
- Aplastic anemia
- Bone marrow infiltrative dis.
- Large granular lymphocyte dis.
- Myelodysplastic syndrome

Anemia: Peripheral Blood Smear Analysis w/ Likely Diagnosis

Schistocytes + Positive	- Microangiopathic hemolytic anemia - Disseminated intravascular coagulopathy - Thrombotic thrombocytopenic purpura - Hemolytic uremic syndrome - Valve Rare Causes: Bongo drummers, marathon runners, Jack hammer operators
Spherocytes	- Hereditary spherocytosis - Toxic envenomations (snake bites) - Clostridial sepsis - Autoimmune hemolytic anemia (AIHA)
Macroovalocytes	- B12-deficiency
Pseudo Pelger-Huet Ano.	- Myelodysplastic syndrome (MDS)
Ferritin is wnl/↑	- Anemia of chronic disease

Hgb electrophoresis: wnl	• Alpha-Thalassemia Minor
Hgb electrophoresis: *Spike **alpha-2 region**	• Beta-Thalassemia Minor
Target Cells [w/ RDW]	• Thalassemia

Chemotherapeutic Drugs: Major Side Effects

5-FU	• Mucositis, diarrhea
Adriamycin (Doxorubicin)	• Cardiomyopathy
Bleomycin	• Lung fibrosis
Cisplatin	• Peripheral neuropathy and Renal failure
Cytoxan	• Hemorrhagic cystitis, Future bladder CA
MTX	• Liver, and lung toxicity
Vinblastine	• Ileus
Vincristine	• Peripheral neuropathy

Coagulation Pearls: ZEBRAS

Factor 13 Deficiency	Factor 12 Deficiency
• Normal PT/PTT and platelet function assay • Patients still bleed	○ Increased ↑PTT time ○ Patients do NOT bleed

Sickle Cell Disease: Clinical Manifestations

Sickle Trait [ST]	• Isosthenuria • Painless hematuria
SC Disease [SC]	• Proliferative retinopathy • Avascular necrosis • Splenic infarcts leading to autosplenectomy
Sickle Cell Disease [SS]	• Secondary hemochromatosis • Parvo B19 • Salmonella osteomyelitis • Pneumococcal infection [Encapsulated organism*] • H. Flu [Encapsulated organism*] • Remember to vaccinate *Pearl: SS can have all the same as SC as well.

Paraneoplastic Syndromes, assoc/w Small Cell Lung Cancer

- Cushing Syndrome
- Lambert-Eaton Syndrome
- Sensory neuropathy and Encephalomyelitis (Hu-Antibodies)
- Syndrome of Inappropriate ADH

Virchow triad

1. Stasis
2. Endothelial damage
3. Hypercoagulability

Triad of risk factors for **Deep venous thrombosis.**

B. HEME/ONC Differential Diagnosis

Anemia of Chronic Disease: DDx

- Atrial myxoma
- Bacterial endocarditis
- Chronic kidney disease*
  *Pearl: CKD most common
- Diabetes mellitus type 2
- Idiopathic pulmonary fibrosis
- Leukemia
- Lymphoma
- Malignancy
- Multiple myeloma
- Osteomyelitis
- Polymyalgia rheumatica
- Renal cell
- Rheumatoid arthritis
- Systemic lupus erythematosus
- Temporal arteritis
- Tuberculosis

Atypical Lymphocytes: DDx

- CMV
- HIV
- Mononucleosis (EBV)
- Cutaneous T cell lymphoma
- Toxoplasmosis

Eosinophilia: DDx

- Allergic bronchopulmonary aspergillosis / Asthma
- Acute interstitial nephritis
- Cholesterol emboli
- Churg-Strauss syndrome
- Drugs
- Eosinophilic pneumonia*
- Loeffler's
- Melioidosis
- Parasites
- *Strongyloides stercoralis*

*Board Buzz Word:** "radiographic negative of pulmonary edema" for **eosinophilic pneumonia***

IgE Elevation: DDx

Allergic Disease	- Allergic rhinitis - Eosinophilic Esophagitis - Asthma - Atopic dermatitis - Food allergy - Drug allergy - Eczema - Urticaria
Drug Effect	- Aztreonam Penicillin G
Infectious Diseases	- Allergic bronchopulmonary aspergillosis - Leprosy
Inflammatory Diseases	- Churg-Strauss syndr. Kawasaki dz.
Malignancy	- IgE Myeloma Hodgkin's lymp.
Parasite worm infestation	- Helminth infection
1° and 2° Immunodeficienc.	- AIDS Hyper IgE syndrome - GVHD Wiskott-Aldrich syndrome

Hemolytic Anemia, Causes: DDx

- Autoimm. hemolytic anemia
- Cold agglutinin disease
- Complement induced lupus
- Drug induced
- Hypersplenism
- IVIG infusion
- Parxosymal cold hemoglob.
- Prosthetic Heart Valves
- Transfusion reactions
- **Direct Trauma:** Marchers, Bongo drummers, marathon runners, Jack hammer operators, burn victims
- **Infection:** Clostridial sepsis, Malaria, Babesia, Bartonella
- **Microangiopathic Hemolytic Anemia:** HUS, DIC, HELLP, TTP, HTN emergency.

Lymphocytic Pleocytosis: DDx

- Acanthamoeba
- Brucellosis
- Cryptococcus neoformans
- Drugs (ex. NSAIDS)
- Fungal
- Leptomeningeal carcinomatosis
- Leptospirosis
- Listeria monocytogenes
- Lymphoma (Facial Nerve Palsy)
- Lyme disease
- Lymphocytic choriomeningitis (ex. hamster exposure)
- Naegleria fowleri
- Sarcoid
- Syphilis
- Tuberculosis
- Vasculitis – Behçet's
- **Partially treated infection**
  - i.e. Prior abx tx/exposure
- **Viral Infection**
  -Enterovirus, Mumps, Arbovirus, West Nile virus, HSV, HIV

Monoclonal Gammopathy: DDx

- Amyloidosis
- B Cell Non-Hodgkin's Lymphoma
- Heavy Chain Disease
- Multiple Myeloma (12%)
- Plasma cell leukemia
- Plasmacytoma
- Smoldering myeloma
- Waldenström's macroglobulinemia

- **Note:** Monoclonal gammopathy of undetermined significance (MGUS), M protein <3g/dL <10% plasma (63% cases), Repeat serum protein electrophoresis 3-6m

Thrombocytosis, Reactive: DDx

- Iron deficiency
- Malignancy
- Chronic infection
- Endocarditis
- Osteomyelitis
- Rheumatoid arthritis
- Tuberculosis

C. HEME/ONC Mnemonics

Anemia, Microcytic vs. Macrocytic

TICS – Microcytic Causes

T	Thalassemia
I	Iron deficiency
C	Chronic disease
S	Sideroblastic anemia

ABCDEF – MACROCYTIC Causes

A	Alcohol + liver disease
A	AZT (Zidovudine - HIV)
A	Agglutination RBCs (Mult. myeloma)
B	B12 deficiency
C	Compensatory reticulocytosis
D	Drug (cytotoxic and AZT)/ Dysplasia (marrow problems)
E	Endocrine (hypothyroidism)
F	Folate deficiency

Disseminated Intravascular Coagulation (DIC)

D	Dx: D dimer
I	Immune complexes
S	Snakebite, shock, heatstroke
S	Systemic lupus erythematosus
E	Eclampsia, HELLP syndrome
M	Massive tissue damage
I	Infections: viral and bacterial
N	Neoplasms
A	Acute promyelocytic leukemia
T	Tumor products*
E	Endotoxins (gram negs, bacterial)
D	Dead fetus (retained)

DISSEMINATED Causes of DIC.

* Tissue Factor (TF) and TF-like factors released by carcinomas of pancreas, prostate, lung, colon, stomach

Eosinophilia

N	Neoplasm
A	Allergic
A	Addison's
C	Connective tissue diseases
P	Parasites

NAACP Main causes for **Eosinophilia**.

Malignancies Likely to Metastasize to Bone

Kinds of tumors leap promptly to bone:

Kinds	Kidney
Of	Ovary
Tumors	Thyroid
Leap	Lung
Promptly	Prostate
To	Testis
Bone	Breast

Ma Pa Pb KTL:

Ma	Pa	P	b	K	T	L
y	a	r	r	i	h	u
e	g	o	e	d	y	n
l	e	s	a	n	r	g
o	t	t	s	e	o	
m	s	a	t	y	i	
a		t			d	
		e				

Pearl:
More than 90% of tumors that metastasize to bone are breast, lung, kidney, thyroid, or prostate.

Target Cells

H	Hemolysis
A	Asplenia
L	Liver disease
T	Thalassemia

HALT
Differential diagnosis of **target cells**

Thrombotic Thrombocytopenic Purpura

F	Fever
A	Anemia (microangiop. hemolytic)
T	Thrombocytopenia
R	Renal failure
U	Unexplained bleeding
N	Neurologic abnormalities

FAT RUN
Clinical picture of **Thrombotic thrombocytopenic purpura**

Thrombocytopenia
ONLY Platelets

V	Viral
I	Idiopathic thrombocytopenic purpura (Immune)
C	Congenital
G.	Gestational
S	Splenomegaly
A	Antiphospholipid syndrome
I	Infection
D	Drugs

2 Cell Lines (Hemoglobin and Platelets)

T	TTP/HUS
E	Evans syndrome
D	DIC

3 Cell Lines (Pancytopenia)

S	Splenomegaly
A	Aplastic anemia
M	Myelofibrosis
M	Myelodysplastic
M	Myeloproliferative

VIC G. SAID to TED and SAMMM

Thrombocytopenia **differential** based on:
1 cell line,
2 cell lines (hemoglobin and platelets),
3 cell lines (pancytopenia)

Chapter 9: Rheumatology

A. RHEUM Clinical Pearls

Antibodies (Serum) for Diagnosis and Treatment	
Serum Antibodies	**Disease**
Anti-mitochondrial	- Primary biliary cirrhosis (PBC)
Anti-nuclear, Anti-Smith Anti-dsDNA	- Systemic lupus erythematous (SLE)
Antihistone	- Drug Induced Lupus
Anti-dsDNA	- Systemic lupus erythematosus (high likelihood of renal involvement)
Anti-CCP	- Rheumatoid Arthritis
C-ANCA	- Wegener's granulomatosis
Centromere	- CREST (**C**alcinosis, **R**aynaud's, **E**sophageal dysfunction, **S**clerodactyly, **T**elangiectasia)
Jo-1	- Polymyositis
Ku	- PM/ Dermatomyositis overlap
La/SSB	- Sjögren's syndrome - Possible neonatal lupus
Mi-2	- Dermatomyositis
P-ANCA	- Churg-Strauss syndrome - Microscopic Polyangiitis
Rheumatoid factor	- Rheumatoid Arthritis
Ro/SSA	- Sjögren's syndrome - Neonatal heart block (Lupus) - Subacute cutaneous lupus
Scl-70 or Topoisomerase	- Scleroderma
U1RNP	- Mixed Connective Tissue Disorder

Antiphospholipid syndrome: Criteria for the Definite Diagnosis

- In order to make the definitive diagnosis of APLS, the patient must meet *at least **one** clinical criteria ---**AND**--- at least **one** laboratory criteria.*

Clinical Criteria	- Vascular thrombosis (arterial, venous, or small-vessel thrombus in any organ)
	- Complication of pregnancy
	  > 1 unexplained fetal death after 10 weeks (nml fetus)
	  > 1 premature birth before 34 weeks (morph. normal fetus)
	  > 3 unexplained consecutive spontaneous abortions before 10 weeks of gestation

Laboratory criteria:	- Anticardiolipin antibodies (positive on two or more occasions, at least six weeks apart) - Lupus anticoagulant antibodies (positive on two or more occasions, at least six weeks apart)

Erythema Nodosum: Causes

- Behçet's
- Drugs (ex. Birth control pills)
- Fungus (ex. Coccidioides, Histoplasmosis)
- Idiopathic
- IBD (ex. Crohn's, UC)
- Pregnancy
- Sarcoidosis
- Streptococcal
- Tuberculosis
- Viral
- Yersinia enterocolitis

Proximal Muscle Weakness, Pearls

- Cardinal sign of **myopathy**
- Dermatomyositis Physical Exam *classic findings*:
    - **Gottron's papules**
    - **Periorbital heliotrope rash**
- Workup
    - ☑ Labs = ↑ **Creatinine Kinase** levels
    - ☑ Diagnose Polymyositis/Dermatomyositis with **muscle biopsy** and **electromyography**.

Systemic Inflammatory Conditions associated with Cardiovascular Disease

SLE	▪ Early CAD ▪ Libman Sachs endocard.	Pericarditis Valvular regurgitation
Rheumatoid Arthritis	▪ CAD ▪ Effusion	Leaflet fibrosis
Ankylosing Spondylitis	▪ Aortic insufficiency ▪ Conduction disease	Proximal aortitis/valvulitis
Systemic sclerosis	▪ Hypertension ▪ Renal crisis	Pulmonary arterial HTN
Takayasu Arteritis	▪ Aortic aneurysms + stenosis ▪ Coronary arteritis	Aortic insufficiency Renovascular HTN
Polyarteritis Nodosa	▪ Cardiomyopathy	
Bechets	▪ Myocarditis ▪ Pulmonary aneurysm	Pericarditis

Uveitis, Associated with Rheumatic Disease

Anterior Uveitis	Posterior Uveitis (Choroid, Retina)
▪ Behcet's ▪ Juvenile Arthritis ▪ Spondyloarthropathies ▪ Sarcoidosis ▪ Wegener granulomatosis	○ Sarcoidosis ○ TB ○ Histoplasmosis ○ Syphilis ○ Lyme disease

Vasculitis

Multisystem Disease → Skin (palpable purpura), Myalgia, Renal, Constitutional, Mononeuritis Multiplex, GI disturbances

Primary Causes	Secondary Causes
- Behçet's syndrome	o Endocarditis
- Churg-Strauss syndrome	o Gonococcus
- Drug toxicity/Poisoning	o Hepatitis B and C
- Giant cell arteritis	o Lyme disease
- Polyarteritis nodosa (PAN)	o Malignancy
- Rheumatoid	o Meningococcus
- Systemic lupus erythematosus	o Rickettsial
- Viral	o Syphilis
- Wegener's syndrome	

B. RHEUM Differential Diagnosis

Arthritis, Monoarticular: DDx

- Degenerative joint disease
- Endocarditis
- Gonococcal
- Gout
- Lyme disease
- Pseudogout
- Reactive/Reiter's syndrome
- Sarcoid
- Septic joint
- Trauma

Arthritis, Polyarticular: DDx

- Endocarditis
- Gonococcal
- Gout
- Lyme disease
- Pseudogout
- Psoriatic
- Reactive arthritis (ex. Campylobacter, Salmonella, Shigella, and Yersinia)
- Reiter's syndrome
- Rheumatoid arthritis
- Sarcoid
- Septic
- Systemic lupus erythematosus
- Vasculitis
- Viral (ex. Coxsackievirus, CMV, Enterovirus, EBV, Hep B, Hep C, Parvovirus, and HIV)

Evan's Syndrome*, Secondary Causes: DDx

Autoimmune	SLE, APLS, Sjogrens
Infections	CMV, EBV, Influenza A, Parvovirus, Hepatitis, Varicella, Nocardia, Lashmaniasis
Malignancy	CLL, B-cell and T-cell non-Hodgkins lymphoma, Plasma cell myeloma, MGUS, Amyloidosis, Kaposi's Sarcoma, CML
Immunodefic.	CVID, IgA deficiency
Other	Grave's disease, dermatomyositis, pregnancy, GBS, UC, BOOP, Castlemans disease, Celiac disease

*Evans syndrome = presence of simultaneous or sequential direct Coombs-positive autoimmune hemolytic anemia (AIHA) in conjunction with immune-mediated thrombocytopenia, with unknown etiology.

Fatigue, Muscular Weakness: DDx

- Muscular dystrophy
- Polymyalgia rheumatica
- Polymyositis/Dermatomyositis
- Sarcoid
- Scleroderma
- Systemic lupus erythematosus
- Thyroid
- Vasculitis
- Myositis from other causes, including medication

C. RHEUM Mnemonics

Cold Agglutinin Disease

M	Mononucleosis [EBV]	5 Ms
M	Mycoplasma pneumonia	
M	CytoMegalovirus	Causes of **cold agglutinin disease.**
M	Listeria monocytogenes	
M	LyMphoma	
	Note: Usually also IgM	

Calcium Pyrophosphate Deposition Disease

H	Hypophosphatemia	5 Hs
H	Hypomagnesemia	
H	Hypothyroid	Causes of **calcium pyrophosphate deposition disease.**
H	Hemochromatosis	
H	Hyperparathyroid	

Lupus

S	Serositis (pleuritis, pericarditis)	SOAP BRAIN MD
O	Oral ulcers (oral, palate most specific)	
A	Arthritis (nonerosive, 2 or more joints)	**Lupus** signs and symptoms.
P	Photosensitivity	
B	Blood (anemias)	
R	Renal (proteinuria or cellular casts)	
AI	ANA Immunologic (ANA>1:160, DSDNA)	
N	Neurologic [psych, seizures]	
M	Malar rash (cheeks & nasal bridge)	
D	Discoid rash (erythematous raised lesions)	

Rheumatoid Arthritis

R	Rheum. factor + 80%/Radial deviation wrist	**Rheumatism**
H	HLA-DR1 and DR-4	
E	ESR/Extra-articular (rest. lung dz, subcut. nod.)	**Rheumatoid arthritis** clinical picture.
U	Ulnar deviation of fingers	
M	Morning stiffness/MCP+PIP joint swelling	
A	Ankylosis/Atlanto–axial joint subluxation//ANA pos 30%	
T	T-cells (CD4)/TNF	
I	Inflammatory synovial tissue (pannus)/IL-1	
S	Swan-neck, Boutonniere, Z-deform- thumb	
M	Muscle wasting (hands)	

D. RHEUM Clinical Case #1

History
- 72-year female with arthritis and recently diagnosed interstitial lung disease. In the course of her work-up, the doctor obtains an ANA level of 1:160.
- → Does she have **Systemic Lupus Erythematosus (SLE)**?

Clinical Pearls ANA Level
- Discovering an ANA level in a well established patient who is elderly and typically has many confounding chronic health problems may be misleading at first. It is important to remember that following factors can cause **false-positive ANA**.

Causes of False-Positive ANA	Other Causes of Positive ANA
- Age - Pregnancy - Rubella - Thyroiditis - Sarcoidosis - Cryoglobulinemia - Liver disease - Mononucleosis - Hepatitis (viral) - Bacterial endocarditis	■ Mixed connective tissue dz ■ Rheumatoid arthritis ■ Scleroderma ■ Sjögren's syndrome

Laboratory and Diagnostic Findings
- Additional work-up revealed patient had a Rheumatoid Factor of 1:80.

Final Diagnosis
- Rheumatoid Arthritis

Case Pearls
This case lends itself to review the clinical pearls for etiologies of a **false positive Anti-Nuclear Antibody** (ANA). Several critical clinical questions are raised in this case and should be considered whenever ANA levels are positive in a patient's workup:

- → Is **Anti-Nuclear Antibody** (ANA) level always suggestive of a diagnosis of SLE?
- → How to manage the patient with this new finding?

Take Home Point
- A positive ANA test by itself is **not** sufficient for a diagnosis of Lupus. Review the following autoantibody tests for Lupus.

Autoantibody Tests	Description
ANA	■ Screening test; sensitivity 95%; not dx.
Anti-dsDNA	■ High specificity; sensitivity only 70%; level is variable based on disease act
Anti-Sm	■ Most specific antibody for SLE; only 30-40% sensitivity

Anti-RNP	Included with anti-Sm, SSA, and SSB in the ENA profile; may indicate mixed connective-tissue disease with overlap SLE, scleroderma, and myositis
Anticardiolipin	- IgG/IgM variants measured with ELISA are among the antiphospholipid antibodies used to screen for antiphospholipid antibody syndrome and pertinent in SLE diagnosis
Anti-histone	- **Drug-induced lupus** ANA antibodies are often of this type (Procainamide, hydralazine; p-ANCA–positive in minocycline-induced drug-induced lupus)

D. RHEUM Clinical Case #2

History

- 70-year old female with persistent low grade fevers, fatigue, malaise, joint stiffness, anorexia, mild weight loss, and shoulder pain.

Laboratory Findings	
	Patient's Results
WBC	7 X 10^3/cu mm
Hb	9 gm/dL
Plt	200 mL
MCV	84 cu µm
RDW	12
ESR	110 mm/hr

Differential Diagnosis for 3 identified conditions based on lab findings:

Normocytic Anemia –DDx

- Acute blood loss
- Anemia of chronic disease
- Anemia of renal insufficiency
- Combined anemia
- Hemolytic anemia

Anemia of Chronic Disease – DDx

- Always remember Uremia or Chronic renal insufficiency
- Bacterial endocarditis
- Cancer in general
- Diabetes mellitus
- Idiopathic pulmonary fibrosis
- Multiple myeloma
- Myxoma
- Osteomyelitis
- Polymyalgia rheumatica
- Renal cell cancer
- Rheumatoid arthritis
- SLE
- Tuberculosis

<u>Note:</u> Many more possibilities, but keep these in mind first.

ESR over 100 DDx

Infections	Malignancy:	Rheumatologic:
• Bacterial endocarditis • Tuberculosis • Osteomyelitis	• Atrial myxoma • Renal cell cancer • Multiple myeloma	• Polymyalgia rheumatica • Temporal arteritis

Case Pearls

- Several clinical pearls contained in this book will be addressed in this case study. Three separate differential diagnosis categories can be used:
    1. **Normocytic Anemia**
    2. **Anemia of Chronic Disease**
    3. **ESR over 100**

D. RHEUM Clinical Case #3

History

- 40-year old female with dry eyes, dry mouth, easy fatigability, and arthritis.

Laboratory Findings	
Na+	132 mEq/l
K+	4 mEq/l
Cl-	110 mEq/l
Bicarb	15 mEq/l
Cr	1.2 mg/dl
SSA	Positive
ANA	Positive

Differential Diagnosis for Non-Gap Acidosis:

- Carbonic Anhydrase Inhibitors (Diamox)
- Fistula (Enterocutaneous, etc.)
- Surgical drains
- Ileal conduit.....Ureteral diversion
- Renal Tubular Acidosis
- Diarrhea
- Sjögren's Syndrome – Renal Tubular Acidosis TYPE 1
- Total Parenteral Nutrition (Amino Acids)

Make the Diagnosis: Tip for Success

- After calculating the anion gap in this case, the differential diagnosis for non-gap acidosis can lead you to the correct diagnosis.

Final Diagnosis

- **Sjögren's Syndrome**

Clinical Summary - Sjögren's Syndrome

- **Sjögren's Syndrome** is characterized by the association of **dry eyes (xerophthalmia)** and **dry mouth (xerostomia)**, with **polyarthritis**. The presence of **sicca (dryness)** symptoms in the absence of another connective tissue disease is designated "**primary Sjögren's Syndrome**," whereas their occurrence in association with another autoimmune process, such as rheumatoid arthritis, systemic lupus erythematosus, progressive systemic sclerosis, or polymyositis, is termed **secondary Sjögren's Syndrome**.
- Source: http://www.hopkins-arthritis.org/arthritis-info/sjogrens/

D. RHEUM Clinical Case #4

History:
- 45 year old female with pruritus and darkening of skin.

Laboratory and Diagnostic Findings:

Patient's Results	
AST	80
ALT	70
ALK Phos	500
TB	3.0

Work-up for Pruritus with Cholestatic Pattern of Liver Injury:
- AMA level
- Liver Function Tests
- Check for Xanthelasma
- Lipids
- Bilirubin

Differential Diagnosis for Pruritus:
- DDX – primary skin vs. primary medical
- Diabetes Mellitus (uncontrolled)
- Hepatitis viral or drug
- Iron deficiency
- Lymphoma
- Mastocytosis
- Polycythemia vera
- Primary biliary cirrhosis
- Primary sclerosing cholangitis
- Thyroidal disease
- Uremia

Final Diagnosis:
- **Primary Biliary Cirrhosis**

Pearls: Primary Biliary Cirrhosis
- Immune-mediated destruction of **small bile ducts.**
- Clinical picture → **Women**, Intermittent **pruritus**, fatigue, and abdominal pain; **jaundice** late in disease, Impaired fat-soluble vitamin (A, D, E, K) absorption leading to osteomalacia, **night blindness**.
- Labs and Diagnostics → **Antimitochondrial antibody (AMA) positive** in most cases. Alkaline phosphatase elevated, aminotransferases typically mildly elevated, liver biopsy reveals portal tract infiltrates and **bile duct injury**. Elevated total cholesterol *without* increased cardiac risk.

D. RHEUM Clinical Case #5

History:
- 30 yo male with cough, arthritis, and red painful bumps on his legs.

Diagnostic Findings:	
Serum Calcium	Calcium = 12
Chest X-Ray	Bilateral Hilar Adenopathy

Final Diagnosis:
- **Sarcoidosis**

Pearls: Hypercalcemia

Major causes of **Hypercalcemia** can be remembered with mnemonic: ***MISHAP***

M	**M**alignancy (Thyroid, Breast, Lung)
I	**I**ntoxication (Vitamin D, Vitamin A)
S	**S**arcoid (RTA – distal) (granulomas)
H	**H**yperparathyroid
A	**A**lkali (Milk Alkali Syndrome)
P	**P**aget's Disease

*Pearl: Other causes of Hypercalcemia are as follows:
- Distal RTA, Pheochromocytoma, Familial Hypocalciuric Hypercalcemia, Addison's, Thyrotoxicosis, Tuberculosis, Immobilization, Lithium, HCTZ

Pearls: Erythema Nodosum

Physical Diagnosis Skill → **discovering painful red bumps is indicative of Erythema Nodosum:**

- Acute, nodular, erythematous eruption usually limited to the extensor aspects of the **lower legs**. The inflammatory reaction occurs in the **panniculus**.
- **To remember the conditions associated with erythema nodosum** use the mnemonic ***SPUD BITS***:

S	**S**treptococcal infection
P	**P**regnancy
U	**U**nknown (idiopathic)
D	**D**rugs
B	**B**ehçet's
I	**I**nflammatory bowel disease
T	**T**uberculosis
S	**S**arcoidosis

D. RHEUM Clinical Case #6

History:
- 42-year old, Caucasian, female admitted to the hospital complaining of knee pain, swelling of the shins, elbows, and ankles. Patient denies difficulty with vision.

Presentation and Examination:
Physical Exam Findings:
- **Skin:** Erythema nodosum of the anterior shins bilaterally

Laboratory Findings:

ACE Levels	WNL
RF, ds-DNA, anti-CCP	WNL
Chest X-Ray	Bilateral Hilar Adeno., Nodular shadows
Mediastinum lymph node biopsy	Noncaseating granulomas

Arthritis, Polyarticular: DDx

- Endocarditis
- Gonococcal
- Gout
- Lyme disease
- Pseudogout
- Psoriatic
- Reactive arthritis (ex. Campylobacter, Salmonella, Shigella, and Yersinia)
- Reiter's syndrome
- Rheumatoid arthritis
- Sarcoid
- Septic
- Systemic lupus erythematosus
- Vasculitis
- Viral (ex. Coxsackievirus, CMV, Enterovirus, EBV, Hep B, Hep C, Parvovirus, and HIV)

Final Diagnosis:
- Löfgren's Syndrome – Acute Onset Sarcoidosis and Polyarthralgia

Clinical Pearls: Löfgren's syndrome
- Commonly, **good prognosis and no treatment warranted.** If refractory musculoskeletal symptoms exist, consider NSAID or prednisolone (recheck ESR and CRP for therapeutic response).
- **Polyarthralgia** and **erythema nodosum** are common in various collagen diseases, including RA. If clinically suspicious for **Lofgren's syndrome**, consider a **biopsy** from skin rash, lung nodules, or **lymph nodes** (as in this case) to reach early/precise diagnosis.

Chapter 10: Dermatology

A. DERM Clinical Pearls

HIV, Dermatological Manifestations

- Aggressive response to insect bites
- Bacillary angiomatosis
- Eosinophilic folliculitis
- Kaposi's sarcoma
- Molluscum contagiosum
- Oral hairy leukoplakia
- Oral thrush
- Scabies
- Seborrhea
- Xerosis

Oral Ulcers: Pearls and DDx

- Aphthous
- Behçet's
- Chemotherapy medications
- Coxsackie
- Herpes simplex virus
- HIV
- Iron deficiency
- Pemphigus
- Reiter's syndrome
- Syphilis
- Systemic lupus erythematosus

Clinical Pearls for Oral Ulcers: To differentiate major oral ulcers focus on **painful vs. nonpainful** presentation of symptoms.

The following cause **PAINFUL** ulcers:

- ☑ **Aphthous**
- ☑ **Behçet's:** Recurrent ulcers located in mouth, genitals, and eyes.
- ☑ **HSV 1 and 2:** Grouped vesicular lesions on lips and oral cavity.
- ☑ **Stevens-Johnson syndrome:** Eroded, bullous lesions on physical exam. Pertinent positive in history of recent medication intake.

Urticaria Workup

- ☑ ANA
- ☑ ESR
- ☑ Food Allergy
- ☑ RF
- ☑ SPEP
- ☑ Viral
- ☑ IgE
- ☑ Parasites
- ☑ Hepatitis profile
- ☑ HIV

B. DERM Differential Diagnosis

Erythroderma, Diffuse: DDx

- Atopic or contact dermatitis
- Cutaneous T cell lymphoma / Mycosis fungoides / Sézary's syndrome
- Drug reaction
- Paraneoplastic / Solid tumor (think lymphoma)
- Psoriasis
- Scabies

Pruritus: DDx

- Diabetes mellitus (uncontrolled)
- Drugs
- Iron deficiency
- Hypothyroid
- Liver failure
- Lymphoma (both Hodgkin's and Non Hodgkin's)
- Mastocytosis
- Myeloproliferative disorders
- Parasites (ex. Scabies)
- Polycythemia vera
- Primary biliary cirrhosis
- Uremia

Rash, Papulosquamous: DDx

- Drug reaction
- HIV
- Mycosis fungoides
- Pityriasis rosea
- Psoriasis
- Syphilis

C. DERM Mnemonics

Erythema nodosum

S	**S**treptococcal infection	**SPUD BITS**
P	**P**regnancy	
U	**U**nknown (idiopathic)	Conditions associated with **erythema nodosum**.
D	**D**rugs	
B	**B**ehçet's	
I	**I**nflammatory bowel disease	
T	**T**uberculosis	
S	**S**arcoidosis	

Melanoma/Malignancy

A	**A**symmetry	**ABCDE**
B	**B**order: (Irregular, scalloped)	
C	**C**olor (>1 color, black, white, blue, red)	**Nevus** that may be a **melanoma / malignancy**.
D	**D**iameter (≥6 mm, pencil eraser)	
E	**E**volving (Increasing in size overtime)	

Ulcers, Genital

S	**S**yphilis	**Some Girls Love Licorice, but Fellows Hate Candy**
G	**G**ranuloma inguinale, and	
L	**L**ymphogranuloma venereum are...	
L	Pain**L**ess	
F	Pain**F**ul ones are...	Distinguishing painful/painless **genital ulcers**.
H	**H**erpes simplex	
C	**C**hancroid	

White Patch on Skin

Vitiligo	**V**itiligo	**Vitiligo PATCH**
P	**P**ityriasis alba/ **P**ost-inflammatory hypopigment **P**iebaldism	Differential diagnosis of **white patch of skin**.
A	**A**ge related hypopigmentation / **A**lbinism / **A**cquired (common)	
T	**T**inea versicolor/ **T**uberous sclerosis (ashleaf macule)	
C	**C**ongenital birthmark (uncommon)	
H	**H**ansen's (leprosy) / **H**ypomelanosis of Ito	

Chapter 11: Pharmacology

A. PHARM Clinical Pearls

Amiodarone: 5 Classic Side Effects

1	Eyes	Corneal deposits
2	Liver	Transaminitis and hepatitis
3	Lungs	Fibrosis and lipoid pneumonitis
4	Skin	Blue-gray skin discoloration
5	Thyroid	Hypothyroidism or Hyperthyroidism

Workup:
- ☑ Check TSH, PFTs, DLCO
- ☑ Caution: Watch for INR interactions (Digoxin, Coumadin)

Anticholinergic Overdose:

- Complete medication history, incl. OTC meds & herbals.
- Drugs to watch out for causing **anticholinergic overdose**:
  - Amitriptyline
  - Atropine
  - Benadryl
  - Hyoscyamine
  - Scopolamine
  - Recreational Drugs
    - Mushrooms, Datura species – *Angel's Trumpet*

Lasix Drip Management

- Lasix Drip → 360mg Lasix: 20/cc/hr
- 480 cc D5W/50gm Albumin

Take Home Message: Used for situations of refractory edema, anasarca, etc. (third spacing). Rarely used.

Salicylate [Aspirin] Toxicity

Signs & Symptoms
- ↑ Prothrombin time (PT), GLU
- Mental status change
- Dehydration and fever
- Noncardiac pulmonary edema
- Promotes renal excretion→bicarb
- Respiratory alkalosis
- Metabolic acidosis

Salicylate [Aspirin] Detox
- ☑ Intravenous fluids and alkalinize urine.
- ☑ If needed, next step → **hemodialysis**.

Warfarin: Drugs that Potentiate Warfarin's Effect

Take Home Message: Patients on Warfarin who are also taking these medications are at risk for increased ↑ bleeding due to the potentiating effects of these meds.

- Amiodarone
- Azoles
- Cimetidine
- Ciprofloxacin
- Erythromycin
- Isoniazid
- Metronidazole
- NSAIDs
- Sulfonamides

Warfarin, Managing the Overcoagulated Patient

Remember: If patient has <u>**active bleeding**</u> administer **Vitamin K, (10mg or IV)** and **fresh frozen plasma**.

INR	Clinical Management
< 6	Hold Warfarin 2-3 days
6-10	0.5 to 1 mg of Vit K, orally
10-20	3 to 5 mg of vitamin K orally
>20	5 to 10 mg of vitamin K, repeat every 12 hours if INR ↑

Warfarin's Effect on Hypercoag Tests

Take Home Message: When a patient is already on Warfarin and he or she need to undergo testing for hypercoagulation, only the following tests marked "OK to test" can be performed and have viable results.

Condition/Test	Testing Viable
Protein C	No
Protein S	No
Factor V Leiden	OK to test Test: APC-resistance assay ----or---- DNA based assay for Factor V Leiden
Antithrombin III deficiency	OK to test Test: Heparin cofactor activity of antithrombin III
Antiphospholipid syndrome	OK to test Test: Clotting based assay to detect Lupus anticoagulants or ELISA to detect antibody against the antigenic complex (cardiolipin and Lupus Anticoagulant)
Hyperhomocysteinemia	OK to test Test: Fasting homocysteine levels.
PT 20210	Ok to test

Warfarin, Diet Guidelines and restriction

Take Home Message: If patient is taking Warfarin these vegetables should be **restricted**.

Vitamin K Containing Veggies	Level of Vitamin K
Broccoli, Brussels sprouts, Cabbage, Spinach, Lettuce, Chickpeas	High
Asparagus, Avocado, Peas, Cauliflower	Medium
Carrot, Celery, Cucumber, Peanut, Pepper, Potato	Low

B. PHARM Mnemonics

Anticholinergic/TCAs/Benadryl Overdose:

Hot as a hare	Fever
Blind as a bat	Blurred Vision
Mad as a hatter	Confusion
Red as a beet	Flushing
Dry as a **bone**	Dry mouth
Bowel and Bladder lose their **tone**	Urinary retention, Constipation
Heart races **alone**	Tachycardia

HOT, Blind, Mad... for **anticholinergic /TCA/Benadryl overdose** signs and symptoms.

Examples:
Anticholinergic
→ scopolamine
Antihistamine
→ diphenhydramine (or Benadryl)
TCA
→ amitriptyline

Chapter 12: General

A. GEN Clinical Pearls

Amyloidosis: Clinical Picture
- Carpal tunnel syndrome
- Congestive heart failure
- Orthostatic hypotension, hyperglobulinemia
- Nephrotic syndrome
- Peripheral neuropathy

Beri Beri: Clinical Picture
- Cardiomyopathy
- ↑High output heart failure
- Neuropathy
- Peripheral edema

Creatinine Phosphokinase (CPK): Causes of Elevation
- Hypothyroidism
- Myocardial Infarction
- Rhabdomyolysis

Edema: Causes of Edema in Lower Extremities

Cardiac	- CHF - Constrictive pericarditis - Cor pulmonale - Hypertension - Pulmonary HTN - Inferior vena cava syn. - ↑High-output fail. - Valvular dysfunction
Liver	↓ Low albumin
Lymphatics	Lymph blockage
Malnutrition	Hypoalbuminemia 2° to malnutrition.
Medications	-Calcium channel blockers -Estrogen -Minoxidil -NSAID -Steroids
Renal	Nephrotic syndrome
Thyroid	Hypothyroidism

Hair Loss: Telogen Effluvium

Causes of Telogen Effluvium	Workup
- Diet Major illness - Pregnancy Stress - Surgery Trauma	☑ Check medication list ☑ CBC, VDRL, ANA, LFT's

Increased Intracranial Pressure (ICP): Non-Struc. Causes
- Accutane (Isotretinoin)
- Cryptococcus
- Estrogen
- Polar bear liver ingestion (Vit A)
- Phenothiazine
- Pseudotumor cerebri
- Steroid withdrawal
- Tetracycline

IV Immune Globulin: Conditions that may warrant administration of IVIG:

- Dermatomyositis/Polymyositis
- Graft vs. Host Disease
- Guillain Barré syndrome
- Chronic demyelinating inflammatory polyneuropathy (CDIP)
- Idiopathic thrombocytopenic purpura
- Severe Group A beta-hemolytic streptococci infections
- Kawasaki disease
- Myasthenia Gravis

Löfgren Syndrome

- Acute sarcoidosis with good prognosis
- Triad of Löfgren syndrome includes:
    1. **Erythema nodosum**
    2. **Bilateral hilar adenopathy on chest radiograph**
    3. **Arthritis**

Overdoses and Antidotes

Toxic Ingestion	Antidote
Acetaminophen	N-Acetylcysteine
Arsenic, mercury	Succimer, dimercaprol
Barbiturates (phenobarbital)	Urine alkalinization, dialysis, activated charcoal
Benzodiazepines	Flumazenil
Beta blockers	Glucagon
Black widow bite	Calcium gluconate, methocarbamol
Carbon Monoxide	100% Oxygen
Cholinesterase inhibitors	Atropine, pralidoxime
Copper or gold	Penicillamine
Cyanide	*Cyanide Antidote Package*: Amyl nitrate + sodium nitrate + Sodium Thiosulfate
Digoxin	Normalize potassium, digoxin antibodies
Heparin	Protamine sulfate
INH	Pyridoxine
Iron	Deferoxamine
Lead	Edetate (EDTA), dimercaprol
Methanol	EtOH, fomepizole, calcium gluconate for ethylene glycol
Methemoglobin	Methylene blue
Muscarinic blockers	Physostigmine
Opioids	Naloxone
Salicylates	Urine alkalinization, dialysis if needed
tPA, Streptokinase	Aminocaproic acid, Cryoprecipitate
Tricyclic Antidepressants	Sodium bicarbonate
Warfarin	Vitamin K, Fresh Frozen Plasma Prothrombin complex concentrate

Sedimentation Rate, Elevated (>100)		
Infections	Malignancy	Rheumatologic
• Bacterial endocarditis • Tuberculosis • Osteomyelitis	• Atrial myxoma • Renal cell cancer • Multiple myeloma	• Polymyalgia rheumatica • Temporal arteritis

Weight Loss: Causes of Unintentional Weight Loss

- COPD
- Depression
- Diabetes (uncontrolled)
- HIV wasting syndrome
- Hyperthyroidism
- Liver disease
- Malignancy
- Tuberculosis
- Uremia

B. GEN Differential Diagnosis

Complement States, Decreased: DDx

- Cholesterol emboli
  - *Purple toe syndrome*
- Endocarditis
- Membranoproliferative GN
- Cryoglobulinemia
- Post-streptococcal GN
- Systemic lupus erythematosus

Weakness: DDx

- Botulism
- Cord problems
- Electrolytes
- Guillain Barré syndrome
- Lyme disease
- Heavy metal poisoning
- Multiple sclerosis
- Myasthenia gravis
- Myositis
- Polymyositis
- Tick paralysis
- Vasculitis

C. GEN Mnemonics

Back Pain

D	**D**egeneration (DJD, Osteoporosis, Spondylosis)	**DISK MASS**
I	**I**nfection (UTI, PID, Potts, Osteo., prostatitis)	Causes for Back Pain
	Injury	
S	**S**pondylitis (Ankylosing spondyloarthropathies, Rheumatoid arthritis, Reiter's)	*See mnemonic about malign. likely to metastasize to bone.
K	**K**idney (renal calculi, infarction, infection)	
M	**M**ultiple myeloma/ **M**etastasis*	
A	**A**bd pain (referred to the back)/ **A**neurysm	
S	**S**kin (Herpes zoster)/ **S**train/ **S**coliosis	
S	**S**lipped disk/ **S**pondylolisthesis	

Pearl: "RED FLAGS" may herald grave underlying condition(s):
- Adenopathy
- Anemia
- Corticosteroids
- EtOH / IV Drug Abuse
- Fever
- Incont. bowel/bladder
- Focal Neuro deficits
- Previous malignancy
- Weight loss

Diagnostic Approach to Challenging Cases

I	**I**atrogenic, **I**nfectious	**I VINDICATE**
V	**V**ascular	Differential diagnosis for **initially unsolved cases.**
I	**I**nflammatory	
N	**N**eoplastic	
D	**D**egenerative/**D**eficiency states	
I	**I**ntoxication	
C	**C**ongenital	
A	**A**llergic/**A**utoimmune	
T	**T**rauma	
E	**E**ndocrine	

Pearl: **I VINDICATE** used in difficult cases when the initial differential diagnosis and lab testing has not yielded an answer. It reminds the clinician to *"go back to the drawing board"* to search out other possibly overlooked causes.

NOTES Section

NOTES Section

NOTES Section

NOTES Section

NOTES Section

www.ingramcontent.com/pod-product-compliance
Lightning Source LLC
Chambersburg PA
CBHW072223170526
45158CB00002BA/730

PATTERNS